SpringerBriefs in Applied Sciences and Technology

SpringerBriefs in Computational Intelligence

Series Editor

Janusz Kacprzyk, Systems Research Institute, Polish Academy of Sciences, Warsaw, Poland

SpringerBriefs in Computational Intelligence are a series of slim high-quality publications encompassing the entire spectrum of Computational Intelligence. Featuring compact volumes of 50 to 125 pages (approximately 20,000-45,000 words), Briefs are shorter than a conventional book but longer than a journal article. Thus Briefs serve as timely, concise tools for students, researchers, and professionals.

More information about this subseries at https://link.springer.com/bookseries/10618

Samiksha Shukla · Jossy P. George · Kapil Tiwari ·
Joseph Varghese Kureethara

Data Ethics and Challenges

 Springer

Samiksha Shukla
Department of Data Science
Christ University
Bengaluru, Karnataka, India

Kapil Tiwari
Christ University
Bengaluru, Karnataka, India

Jossy P. George
Christ University
Bengaluru, Karnataka, India

Joseph Varghese Kureethara
Centre for Research
Christ University
Bengaluru, Karnataka, India

ISSN 2191-530X   ISSN 2191-5318   (electronic)
SpringerBriefs in Applied Sciences and Technology
ISSN 2625-3704   ISSN 2625-3712   (electronic)
SpringerBriefs in Computational Intelligence
ISBN 978-981-19-0751-7   ISBN 978-981-19-0752-4   (eBook)
https://doi.org/10.1007/978-981-19-0752-4

This Springer imprint is published by the registered company Springer Nature Singapore Pte Ltd.
The registered company address is: 152 Beach Road, #21-01/04 Gateway East, Singapore 189721, Singapore

Preface

Data is the essential element of contemporary society. This generation has evolved into a data-informed society. Advancements in technology in various forms catapult this evolution. Organizations utilize data for better decision-making, customer-centric facilities, and target marketing. Data repurposing is another emerging field, mainly used in the medical/pharmaceutical domain. With the emergence of data at a faster rate, the following quote has become appropriate over time, **"Data is the new Soil, Let us cultivate it."** It is a buzzword and a fizzy word these days. The internet is becoming a data facilitator not only for researchers but also for home-makers, kids, and businesses. It is becoming a critical part of our lives in a growing data-driven society. The Internet of Things has changed the world forever. Research, Academia, Government and Private sectors, Medical institutes, Cultural heritage, meteorological institutions, etc., generate, utilize and share a huge amount of data. This massive use of data comes with many challenges for individuals, institutions, societies, and states.

Considering the widespread relevance of the data today, it is easy to understand how precious it is. Data-driven decision-making provides an ease to individuals, but there is a need to maintain privacy, confidentiality, and security of the data. Ethical data handling practices play a crucial role in a data-informed society.

Data pertaining to a living being is primarily a property of that being. Hence, proper permission is required for anyone else to use that private property. Some vital questions are: Has a living being got absolute authority over its data? Can a living being use its data? What is the right of others over the data of a person? Has the state got any authority over the data of a living being?

Living beings can generally be classified into plants, animals, and human beings. Although it is not a proper biological taxonomy, this classification is based on the activities and the nature of domestication of living beings. Data related to plants and animals have to be collected and used concerning the state's laws. However, data pertaining to human beings have to be treated respecting the person from whom the data is collected and the state laws that govern it. In various cases, the dependent and independent human beings have to be treated separately as far as data collection and data use are concerned.

Religions and social settings have played essential roles in determining the foundations of ethics. The foundation of ethics laid upon the 'good' and 'bad,' and 'right' and 'wrong.' The perspectives are absolute as well as contextual. Ethics has to be thought about as evolutionary. It is because several concerns in ethics have been replaced and reworked in the various stages of history and context. The theories such as teleological, axiological, deontological, formalistic, naturalistic, intuitionistic, non-cognitive, etc., are to be considered in any formal discussion on ethics.

In close association with the data comes the intellectual property of living organisms. One question is important: Should intelligence be restricted to natural intelligence? How much authority does one person have over one's intellect? How much does one person owe to the state and society for one's intelligence? When considering the technical term 'Intellectual Property,' it primarily refers to an individual's 'creative outputs.' Has an individual got absolute authority over one's creative outputs? International and national regulations and directives on intellectual property and intellectual property rights from multiple perspectives have to be discussed.

What to Expect from This Book

This book gives a thorough and systematic introduction to Data, Data Sources, Dimensions of Data, Privacy, and Security Challenges associated with Data, Ethics, Laws, IPR Copyright, and Technology Law. This book will help students, scholars, and practitioners to understand the challenges while dealing with data and its ethical and legal aspects. The book focuses on emerging issues while working with the Data.

Who This Book is For

This book is primarily aimed at the general public, who need to know about multiple aspects of dealing with the data. Particularly, this book comes as a ready-to-refer short handbook for the people involved in the data-related industry, viz., data scientists. As an emerging field of higher education, data science programmes can use this book as a reference book.

Road Map

- Chapter 1 covers the concept of Data, its type, and sources. It also introduces the various facets of the data, the science of data, data ownership, and the FAIR data principles (Findable, Accessible, Interoperable, and Reusable). The importance of use/abuse and overuse of the data. In the current era of the information world of faster data harvesting, data theft and misuse of the data is a critical concern.

- Chapter 2 of this book underlines what data privacy entails and its relevance and importance to individuals in a growing digitized world. The right to privacy and the ethics concerning data collection from individuals is discussed nuanced, tracing constitutional recognitions and governmental rulings of an individual's privacy as a fundamental right. The chapter also presents the legal disputes and cases concerning data privacy as instances to support the information furnished. The chapter thus comprehensively explores the intricacy of data privacy and its evolution.
- Chapter 3 focuses on the security aspect associated with the data. This chapter presents various types of security breaches and mechanisms to control them. The Data Security and Information Technology Act is also covered in this chapter.
- Chapter 4 looks at the ethical aspect associated with data and data science. The concept of data ownership is presented in this chapter. The importance of Informed Consent and Institutional Review Board (IRB) along with repurposing and research ethics are also covered.
- Chapter 5 presents the concept of Intellectual Property Rights (IPR) and Copyright Law.

Bengaluru, India   Samiksha Shukla
                                                                           Jossy P. George
                                                                           Kapil Tiwari
                                                                Joseph Varghese Kureethara

Contents

About the Authors

Dr. Samiksha Shukla is currently employed as Associate Professor and Head, Data Science Department, Christ University, India. Her research interests include computation security, artificial intelligence, machine learning, data science, and big data. She is certified AWS Educator. She has presented and published several research papers in reputed journals and conferences. She has co-edited books such as *Data Science and Security* (Springer, 2020 and 2021). She has 16 years of academic and research experience. She is serving as Reviewer for the *Inderscience* Journal, Springer Nature's *International Journal of Systems Assurance Engineering and Management* (IJSA), and IEEE and ACM conferences.

She is an experienced and focused teacher who is committed to promoting the education and well-being of students. She is passionate about innovation and good practices in teaching. She has been constantly engaged in continuous learning to broaden her knowledge and experience. Her core expertise lies in computational security, artificial intelligence, and healthcare-related projects. She is skilled at adopting a pragmatic approach in improvising on solutions and resolving complex research problems. Dr. Shukla possesses an integrated set of competencies that encompass areas related to teaching, mentoring, strategic management, and establishing a centre of excellence via industry tie-ups. She has a track record of driving unprecedented research and development projects with international collaboration and has been instrumental in organizing various national and international events.

Dr. Jossy P. George is currently serving as Dean and Director at Christ University, India. He has a dual master's degree in Computer Science and Human Resources from the USA and has done his FDPM from IIM, Ahmedabad. He was awarded a Doctorate in Computer Science by Christ University, India, on Image Processing for the paper titled 'Development of Efficient Biometric Recognition Algorithms based on Fingerprint and Face', and his research activities continue to focus on algorithms for improved accuracy in biometrics. He has also completed major research work on the topic of 'Technology in Higher Education'. He has published many papers both in national and international journals and has also successfully filed for patent applications. In addition to this, he is Author of two motivational management books.

As Educator, he has been working with Christ University (previously Christ College), Bengaluru, and other associated institutions since 2002 in various capacities. In his current capacity as Dean and Director, he continues to strive towards CHRIST's motto of excellence and service. He resolutely believes in a holistic approach to education and is motivated by this to encourage the students of his institution to excel at both academic and non-academic pursuits.

Mr. Kapil Tiwari is currently working with Amazon as Software Development Manager. He has earlier worked with OpenText, EMC, Sun Microsystems, SS&C, and Siemens information system Ltd. He has M.Sc. in Information Technology from Devi Ahilya University, and he is an alumni of C-DAC ACTS, Bangalore. He has 17 years of professional experience in managing, designing, and developing cloud and on-premise software systems. He also has rich exposure to finance technologies, content management, and the medical domain. He is currently pursuing Ph.D. from Christ University, India, in the field of privacy in machine learning.

Dr. Joseph Varghese Kureethara is heading the Centre for Research at Christ University. He has over seventeen years of experience in teaching and research at Christ University, India, and has published over 100 articles in the fields of graph theory, number theory, fuzzy optimization, history, religious studies, and sports both in English and Malayalam, in journals such as *International Journal of Fuzzy Systems*, *Journal of Cleaner Production*, *IET Renewable Power Generation*, *International Journal of Hydrogen Energy*, *Socio-Economic Planning Sciences*, and *Match*.

He has authored three books and co-edited six books including *Recent Trends in Signal and Image Processing* (Springer, 2021), *Data Science and Security* (Springer, 2021), and *Neuro-Systemic Applications in Learning* (Springer, 2021). His blog articles, comments, facts, and poems have earned about 1.5 lakhs total page views. He has delivered invited talks in over thirty conferences and workshops. He is Mathematics Section Editor of *Mapana Journal of Sciences*, Member of the Editorial Board, and Reviewer of several journals. He has worked as Member of the Board of Studies, Board of Examiners, and Management Committee of several institutions. He has supervised five PhDs, 12 MPhils, and is supervising seven PhDs and is employed as Professor of Mathematics, Christ University, Bengaluru, India.

Chapter 1
Data and Its Dimensions

In God we trust. All others must bring data.—W. Edwards
Deming

Data is a collection of raw facts. Data is all around us, and we are generating data at speed never seen before. Everything we do over the internet creates data and adds to the data pool. Data is one of the essential elements of society these days. The technology changes have led to datafication. Datafication is a technological transformation to meaningful information which gives a new form of value to the data. It is the need for data science in all aspect of life, which make it omnipresent. Every aspect of life is datafied from the books we read to film we watch and the food we eat, the individual physical activity, our purchases, our social media presence, etc.

While collecting data, one of the significant aspects is informed consent. **Informed Consent** means the purpose of the data collection should be informed to the population from whom the data is collected. Informed Consent gives researchers the freedom to use data without any legal challenges (see Chap. 4).

The structure of this chapter is as follows. In Sect. 1.1, there is a detailed discussion on the two types of data. In Sect. 1.2, the Theory of information is presented. Sources of Data are presented in Sect. 1.3. Section 1.5 offers the Science of Data. The next section briefly discusses the idea of Data Ownership. Sections 1.7 and 1.8 are about the value of data and use abuse, and overuse of data, in Sect. 1.9, Data and Information Technology Act. Section 1.10 presents a popular Cambridge Analytica and Facebook.

1.1 Types of Data

Data is a collection of facts [1], but it can also mean words, sounds, images, or simple descriptions. It can be classified into two categories; Fig. 1.1 shows the representation of data types:

S. Shukla et al., *Data Ethics and Challenges*, SpringerBriefs
in Computational Intelligence, https://doi.org/10.1007/978-981-19-0752-4_1

Fig. 1.1 Types of data

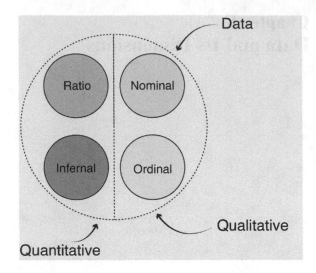

- Qualitative Data
- Quantitative Data

Qualitative Data

Qualitative Data is non-numeric, which is observed to be subjective and descriptive. It is also defined as the data that estimates and characterizes, which can be monitored and recorded. Qualitative data can be gathered through one-to-one interviews, observations, focus groups meeting, etc. It can be grouped based on the categories, which is also known as categorical data. In Statistics, categorical data is the data that can be arranged unconditionally based on the features and dimensions of things.

There are three classifications of categorical variables: binary, nominal and ordinal. A categorical data which can take only two possible states, such as True/False, Yes/No, is treated as **binary variables**, for ex. Win/Loss in a game, Heads/Tails in a coin flip. A **nominal variable** is labeled data classified into various groups with no ranks or order between them, for example, country, gender, or hair color of a group of people. An ordinal variable is grouped based on order or ranking, for example, the Economic status of an individual as 'wealthy,' 'middle class,' and 'poor.' An ordinal variable can also be used as a quantitative variable if the scale is numeric.

For example, think of a child reading a paragraph from a textbook while studying at home. The mother listening to the reading gives feedback on how the child read that paragraph. The mother provides feedback based on fluency, pitch, clarity in pronunciation without giving any grade; this is considered an example of qualitative data.

- The ice cream is yellow, pink, and purple (qualitative).
- Humans have brown, black, blue, and red eyes (qualitative).

Quantitative Data

Quantitative Data is defined as the value of data in numbers. Here, every data set has a unique numerical value associated with it. It is quantifiable and can be treated and manipulated using mathematical calculations and statistical analysis. Quantitative data answers questions such as "How much?", "How often?" and "How many?". This data can be verified and conveniently evaluated using a mathematical approach.

For instance, "How much did that necklace cost?" queries to collect quantitative data. Quantities correspond to various parameters. Values are associated with every measuring parameter such as Kilogram or Pound for weight, Dollar or Rupees for cost, Liters for liquid, etc.

Quantitative data provide ease in measuring different parameters. Generally, it is collected for analysis using questionnaires, surveys, or polls circulated across to a particular section of a population. The results obtained can be applied across the population.

- How much was the price of that car?
- What is your age?

Need of the Data

Data is a ubiquitous resource that can provide shape to innovations and insights. But if we look at it with unaided sight, it is just a pile of numbers and isolated facts. However, exciting results emerge if you start working with it and playing with it in specific ways. Data is a goldmine; let us dig into it; these days, data is generated at a tremendous rate. The value of data is known to every individual; as we dig into it, we get more information and understanding. The insight gathered from the data is utilized by the company for business enhancement and better decision-making.

In one of his TED talks, *"Data is the new Soil,"* David McCandless says that we have to cultivate data. It is a fertile creative medium. It is significant in the current scenario. Over the years, we have contributed a massive amount of data online. The data is irrigated from the network. We are dependent on data for our every move, such as buying a new car or air conditioner for summer. Users today take opinions based on the existing data, such as sales data, users' feedback, etc.

Data is required to make informed decisions. It is needed that data should be explored for better decision-making. Understanding the data helps businesses to move forward in a positive direction. Good data brings undeniable facts, while anecdotal facts, presumption, or abstract observation might lead to wasted resources as the decision was based on a wrong conclusion.

Data is needed to ease life for people around you and help them make better decisions. "Quality is the main reason which must be considered by the organizations that use data."

Data is needed to monitor and maintain the health of an organization's critical systems. With the help of data analysis, organizations can respond to the challenges before it becomes unmanageable. Adequate quality tracking/supervision will allow organizations to handle challenges proactively. It facilitates organizations to retain best practices (Fig. 1.2).

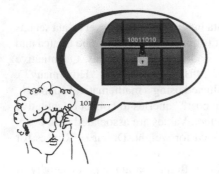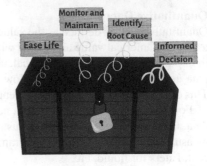

Fig. 1.2 Need of the data

Data can be collected to weigh the effectiveness of the selected strategy. It gives us an idea of how well the strategic solution is performing. It can also enable us to decide whether or not to tweak the strategy for better performance. The vendors use this strategy during the sale announcement; the offers are brought in the market, and based on customers' responses, the strategy is weighed to decide the next action plan.

Data can be used to identify the root cause of the problems. Based on the reason recognized, a solution can be built. If the data is considered parallelly, it can result in more accurate solutions.

The decisions supported with the data have more acceptance than decisions without any data backup. In the legal field also, we can see the importance of the evidence to announce the judgment.

Data analysis can help organizations to identify the area of strengths and weaknesses. High-performing employees, programs, and projects can be identified with the help of data analysis. The data is a critical factor in moving forward in the right direction and setting benchmarks and goals.

1.2 Theory of Information

The Theory of Information is the scientific study of information storage, quantification, and communication. It is the proper merging of Probability, Statistics, and Computer Science and has made significant contributions to Deep Learning and Artificial Intelligence. It can be seen as a sophisticated amalgamation of basic building blocks of deep learning, calculus, probability, and statistics. It is the mathematical treatment of the concepts, parameters, and rules governing the transmission of messages through communication systems [2].

1.3 Sources of the Data

The means of collecting data is termed a Data Source. The data sources can be classified into two main categories: (1) Primary Sources (2) Secondary Sources.

Primary and Secondary Sources of Data
Primary data sources provide the data collected by researchers directly through interviews, surveys, etc. It is the first-hand report of the event approved by officials. It is a representation of original thinking, documentation of the research findings. These sources are mainly created at the time of occurrence of the event, but they can also include references created later.

Primary data sources include original documents like your experiments, social security number, aadhar cards, passport cards, birth certificates, ship logs, biographies, autobiographies, constitutions, statistical data, literature, etc.

Secondary Sources
Secondary data sources present evaluations, interpretations, or analyses of primary sources; these include Census data collected by the government, textbooks, book reviews, encyclopedias, dissertations, newspapers, and political commentary.

The secondary source of data is the data created later by the one who did not participate in the event and has no first-hand experience of the work. It includes evaluation, analysis, or interpretation of first-hand information.

Primary and secondary sources are usually not fixed; it depends upon the study. For example, a journal article can be both primary and secondary. If it discusses the state-of-the-art technological innovation, it is primary, but if it is considered a reference of the study, it is secondary.

Data from Person or Things
The data can be classified as data about a person or things. Most of the time, the data is related to humans, i.e., data about humans, by humans, or for humans. It is the data about human beings, animals, and things. The data about humans help in many aspects of life, such as disease data to help the medical professionals to give better treatment, the data about the food preference in a region, etc. The data about animals and plants provides detailed information about the animal and plant species. The New York State Department of Environmental Conservation collects and stores data about the species of the plants and animals. One such data under the biodiversity program is the data about the Rare animals and Rare Plants. These days, the data about things is widely used, such as recommendation systems or review and feedback data about some product. Most of us are used to checking the features, reviews, and ratings based on the data available about the product or services. These types of data contribute to the well-being of individuals, organizations, and the environment.

Types of the Internet of Things Data Collection
In the current scenario of the smart world, the data is collected through smart devices. In the IoT data collection process, the sensors are used to track the conditions of

material things. These connected devices can measure data in real-time. This data can be transmitted, stored, and fetched at any time.

The IoT data is relevant because, as per the International Data Corporation (IDC) forecast, there are already more connected things than people in the world. It is predicted that there will be 41.6 billion connected devices by 2025. People are keen to go for smart devices and smart homes that are increasing data and the sources of data. With the help of these connected devices, companies are widening the range of data capture. Based on studies, the IoT data can be defined as follows:

- Equipment Data
- Environmental Data
- Submeter Data

Equipment Data

Equipment Data is mainly used to maintain the health of the equipment in the world where the population is moving forward to the Internet of Everything (IoE). With the invention of IoE, connected devices are rising significantly. The data generated through these devices help predict the maintenance timeline, real-time fault detection, planning, and better decision-making. Thus, it helps reduce the cost and energy consumption, increase efficiency, and prolong the life of the equipment. The equipment data is mainly centralized, slowly adapting the decentralized model. The decentralized model comes with some challenges related to privacy and security. The authentication and agreement should be adapted in the decentralized model to avoid theft and spoofing.

Environmental Data

IoT sensors are used for smart homes, medical instruments, and weather monitoring to track temperature, air quality, monitor and manage crowds across different gatherings. The environmental data can be used for alarming and avoiding disasters such as floods, forest fires, and the spread of disease by monitoring the crowd gathering. Many governments and non-governmental organizations such as Climate Change Knowledge Portal (CCKP), NASA's EARTHDATA, National Institute of Standards and Technology Data Gateway (NIST) give access to NIST scientific and technical data.

Submeter Data

These data are collected from the digital submetering devices automated to the internet. It reduces manual errors and is quick. It is beneficial in systematic organization and conservation of the resources by introducing continuous detailed reporting. It can help us get insights about hourly, weekly, monthly, and seasonal utility-reports about resources and equipment. Many factories use water submetering to predict daily water demand.

Social Media Data

Social Media Data is the information a social media user shares publicly. It includes information such as location, link shared, language, etc. Social media data is vast, with billions of shares, likes, tweets, stories, posts, photos, and videos getting added every day. It is collected from the social media activity of individuals. The click metadata helps marketers understand the individual interest and send personalized recommendations. Linkedin, Netflix, Twitter, Facebook, etc., are the most significant contributors to social media data.

Spatial Data

Spatial data/Geospatial data is the data that refers to the geographical location. It includes the data beyond the site's coordinates, such as shape and boundaries. The spatial data can be represented in the form of point, line, or polygon called vector (object) or in continuous cells raster (image) mode. The vector data consist of points, lines, arcs, and polygons with start and endpoints; roads, trees, ponds, cadastral are examples of vector data. Raster Data is represented as a data matrix where each cell represents a geographic location; it is categorized into discrete raster data like population density and continuous raster data such as temperature and elevation. The spatial data sources are analog maps, aerial photographs, satellite images.

Medical Data

Medical data is widely used for research and development purposes. Every visit to a doctor generates a lot of data that can be analyzed. It is data related to the individual patient's health, such as clinical reports and medical history. Medical data analysis can be utilized for better treatment and made available to prevent diseases. The medical data is referred to as Electronic Health Records (EHR). EHR is a digital form of a patient's medical information and history maintained by the physician throughout treatment. It mainly contains **generic information** like temperature, heart rate, blood pressure, **diagnostic information** such as blood tests, culture tests, kidney function tests, imagery like CT scans, X-Rays, and **treatment information** like previous medical history, ongoing medication, and doses. Other clinical information could be administrative data, patient disease registry, health surveys, insurance claims data.

Big Data

Big data is the same as small data but bigger. It can be structured, semi-structured, or unstructured. It contains a huge amount of data that can be computationally analyzed to reveal patterns, get trends and insights from massive data sets, which is practically impossible with traditional data processing applications. It deals with different, voluminous data generated every microsecond [3]. It mainly deals with seven V's, as shown in Fig. 1.3.

Volume

In the 1970s and 80s, only text data were stored and processed by enterprises. With the exponential data growth in the last two years, we have data from various sources in multiple formats. Enterprises maintain petabytes and terabytes of storage; for example, Over one million customer transactions are recorded by Wal-Mart per hour, generating more than 2.5 petabytes of data.

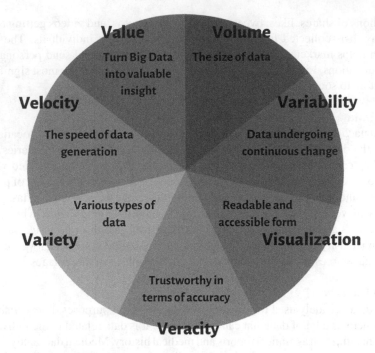

Fig. 1.3 Characteristics of big data

Variety

The variety of data can be described in two dimensions first based on the types of *data formats*; these include a long list of TXT files, video, audio, emails, XML files, CSV files, social media text messages, images, and graphics. Second, based on *the sources,* Variety is also considered the data coming from different sources such as ERP, SCM, POS, Facebook, Twitter, Sensors, Mobile Phones, etc.

Velocity

Velocity refers to the speed of data generation and frequency of delivery. The data flow is massive and continuous, valuable to researchers and businesses for strategic competitive advantages. Highest velocity data typically streams directly into memory versus being written into a hard disk. For processing data with high velocity, tools for streaming analytics were introduced, for example, KAFKA, STORM, etc.

The velocity of data can be described in three dimensions: *Data-in-Motion:* velocity can be described as data-in-movement. For example, the stream of readings taken from a sensor, the clicks by each visitor to a website, readings from a smart meter, etc. *The lifetime of Data Utility:* The second dimension of velocity deals with data validity. Is the data permanently valuable, or does it rapidly age and lose its meaning and importance. *Speed of Processing:* The last dimension of velocity is the speed at which the data must be stored and retrieved.

Value
The huge volume of data does not make sense if it has no value. The data should reflect the purpose for which it is collected, the scenario, and the business outcome it will deliver. If data does not have a value associated with it, it does not worth storing and maintaining. The usefulness (value) of the data makes it worth storing and analyzing to give insight into the data.

Veracity
Veracity refers to the biases, noises, and abnormalities in data. This is to ensure that data is clean and accurate. Here, we identify the relevance of data and ensure data cleansing is performed to store valuable data. It verifies that the data is appropriate for its intended purpose and usable within the analytic model. The data is to be verified against a set of defined criteria.

Variability
Variability in big data is referred to in three dimensions: first is the number of inconsistencies in the data. The second dimension is a multitude of data dimensions from multiple data types and sources. The third dimension is the variable speed at which big data is loaded in the database.

Visualization
Data visualization makes it easy for non-technical users. When represented in graphs, the complex data communicate the data story in common users' perspectives. It transforms the data into information, information into insight, insight into knowledge, and the knowledge can be utilized to make business decisions.

1.4 Facets of Data

Due to the growth of internet technology and tools, there are various data dimensions. In the last two years, we have become rich. The data is growing exponentially, bringing many challenges to handle different types of data, massive size, and the speed generated.

The main types of data are:

- Structured
- Unstructured
- Semi-structured
- Machine-generated
- Natural language
- Graph-based
- Audio, video, and images
- Streaming

Structured Data

Structured Data is the data that has fixed attributes within a record. It is easy to store structured data in relations/tables within the Databases and Excel files. Structured Query Language is the preferred language to manage and query data from the databases.

For example, structured data include dates, phone numbers, names, etc. However, in the current digitization scenario, unstructured data is often generated.

Unstructured Data

Unstructured data is difficult to fit into a data model because the contents are not organized as per the pre-set schema—a very widely used example of unstructured data is social media data. The thousands of different languages and dialects out there further complicate this. Surveillance data, weblog, text files, Geospatial Data, Multimedia data are examples of unstructured data.

In the initial years of computerization, data storage followed a well-defined schema. The changes to this data were complex, but unstructured data took over; many recent applications, such as IoT, mobile activity, and social media use unstructured data.

Semi-structured Data

Semi-structured data does not exist in a relational database, but it has some specific attributes that make it simpler to analyze. With the proper techniques, one can keep them in the relational database. There is no tabular structure as in relational databases. Examples of semi-structured data could be XML and other markup languages, zipped files, weblogs, Email, etc.

Machine-Generated Data

Machine-generated data is automatically generated by a computer, IoT devices, Surveillance Cameras, Applications, or other machines without human intervention. It is becoming a significant data resource and will remain so. Machine data analysis is mainly dependent on scalable tools to match the volume and the speed. Examples of machine data include call details records, network event logs, telemetry data, etc.

As mentioned earlier, International Data Corporation (IDC) predicted that there would be **41.6 billion** connected IoT devices, or "things," generating 79.4 zettabytes (ZB) of data in 2025. They project that the amount of data created by these connected IoT devices will report a compound annual growth rate (CAGR) of 28.7% over the 2018–2025 forecast period. A significant contribution to this data is video surveillance applications; however other classes such as medical and industrial data will contribute more over time.

Natural Language

Natural language is a special type of unstructured data. It is challenging to process Natural Language Processing (NLP) data because it needs knowledge of linguistics and data science techniques to interpret it [4].

The NLP community has been thriving in text recognition, entity recognition, topic recognition, text completion, and sentiment analysis, but the main challenge is the model trained in one domain cannot be generalized well to other domains. It is difficult to recognize every text with some state of art techniques. However, it should not be a major concern as sometimes humans struggle with Natural Language because of its ambiguous nature. The same conversation can be interpreted differently by two people, and the meaning of the same word can vary based on the listener's mood. The idea of meaning itself is cryptic here.

Everything expressed in writing or orally can be interpreted using NLP techniques; it carries enormous amounts of information. It is about the contents and the topic selected and the tone and word selection, which gives us a complete picture of the individual and the thing. This information can even be utilized to predict human behavior.

NLP is becoming a supporting technique in the medical domain by providing improved care, disease diagnosis, and better treatment at patients' comfort with reduced cost.

Graph-Based Data

Graph-based data here is the data concerning the concept of graph theory. The graph theory is a mathematical model representing the pairwise relationship between the nodes. It mainly focuses on the relationship and adjacency of the nodes. Nodes and edges characterize the graph structure. It is primarily used to represent social networks, finding the shortest path. For example, you can see the connection between your friends and the individual nodes with the organization they belong to. The follower list on Instagram is another example of graph-based data. The leading utility of Graph databases can be seen in which overlap can be demonstrated, such as the connection on Facebook, Linkedin can tell us the company profile. Imagine another graph considering the movie preferences on Amazon Prime.

Audio, Image, and Video Data

Due to easy, fast, and cheap access to technology and hardware products, there is a humongous amount of data available in the form of audio, images, and video. Everyone is equipped with Smart devices, making them one click away from contributing to the multimedia data pool. The image recognition task is less challenging for humans but turns out to be difficult for machines. A high-speed camera captures all the games for sports analytics. These cameras can capture each ball and player's movements to calculate in real-time.

DeepMind successfully created an algorithm capable of learning how to play video games. The company prompted Google to buy its Artificial Intelligence (AI) development ideas.

Streaming Data

The streaming data can also be called live data. It is given to the system instead of loading when the event is happening in the data store. It can be generated from different sources and processed incrementally using stream processing techniques. It is adopted when we need the analytics results in real-time. The stream data processing of a transaction is beneficial in fraud detection. In case the transaction data is stream-processed, based on anomalies, the fraud can be signaled/warned in real-time so that the transaction can be stopped before it is committed. Examples of real-time stream data include real-time trending data on social networks like Twitter, Linkedin, etc.

1.5 The Science of Data

Now, as the Data, its types, and the data sources are known. It is the right time to explore the science of data. The word Science is derived from the Latin word *Scientia*, meaning "Knowledge." It is a systematic enterprise that builds and organizes knowledge in testable explanations and predictions about the universe. Data plays an essential role in various domains such as Transport, Healthcare, Weather Forecasting, Inventory Problems, etc. The Science of Data is the systematic study of data to create knowledge. This knowledge can be used for better decision-making. In the current scenario with data hype, individuals are habitual in getting the information at their fingertips. In the pandemic, citizens are educated with data-driven decision-making. The systematic study of data reveals that the prediction of the results is appropriate.

1.6 Data Ownership

Ownership means having complete control over the data, property, an asset, object, or real estate. Data Ownership is termed possession and responsibility of the data/information. It's the power to control creation, insertion, update, deletion, and restrict data access. It is a data governance process that clarifies the organization's legal ownership of the data. The data owner can create, update, share and restrict access to the data. The owner also has the ability to surrender the ownership to a third party [5].

Theft of Data
In this age of information, technology, and experience, data has become a much more valuable asset, with anything created, stored, and shared electronically through smart devices or over the internet. Data has become the instrument through which companies acquire clients and take over a considerable part of the market. Data theft is an act of stealing, deleting, or copying an individual's data or company's confidential data without authorization. The crimes like stealing customer data, identity theft, usage of research outcomes of others without the inventor's permission or acknowledgment are at the hike. An example of a growing trend in identity theft is social media identity theft; on social networking sites, people are encouraged to share their personal information, which puts them at risk of becoming victims of identity theft. Some scammers can use the personal data of individuals to create fake Facebook,

Twitter, Linkedin, or Instagram accounts with your personal information and image. There are other forms of data theft such as credit card information, ATM information, Bank account information, data-stealing from external devices.

Section 43(b) of the IT (Amended) act, 2008 states that an individual would be responsible for data theft if he retains stolen computer resources, copies, downloads, or retrieves any data, or the data stored in the removable storage medium without permission of the owner or any authorized person.

Data can be misused, copied, or hacked. The main risk to data is the employees who have access to it or the third party maintaining it. These data criminals usually sell the data to the competitors to attract customers to their business by putting in less effort and costs.

1.7 Value of Data

The data has some value associated with it. It may not be the price we pay; instead, it is beyond the expected value. Data value is decided based on the scope of utilization and overall benefit to humankind/society. These days data is a moving force for businesses. The value of data for online shopping services such as Amazon, eBay, and Flipkart are not concerned with money or pricing, but preferably to deal with the target personalized marketing, operation optimization, and customer satisfaction.

1.8 Use/Misuse/Overuse of Data

The data is generated at an enormous rate, which the organization uses to establish standards and benchmarks to take the business forward. *At present, the use of data is at its peak,* be it individual or an organization. Everyone contributes to the data pool and tries to get insight from it. For example, the utilization of smartphones for checking the status of pandemic-affected cases in the surrounding locality, E-Commerce companies like eBay, Amazon, mobile application-based services like Uber, Ola, Zomato, Accuweather, and some government applications like CDC's Milestone Tracker, USDA FoodKeeper, Aarogya Setu, ePathshala app collects information for personalized recommendations, target marketing, customer satisfaction.

Data can be classified into the categories-descriptive, diagnostic, prescriptive, and predictive based on usage. The data can describe the analysis for better understanding and decision-making. The knowledge of data can be diagnosed to find the root cause of the problem. Based on the diagnosis, the prediction can prescribe the next action.

Misuse of Data
As the amount of data increases, it comes with many challenges. Data misuse is a critical problem impacting organizations and individuals. It is the use of data in

unexpected ways. Many corporate policies, user agreements, and privacy laws exist to control the collection and use of the data. These policies help to govern data theft and abuse.

For example, a hospital collects patients' data and shares it with an insurance policy agent. It is considered as the misuse of data as it is not utilized for the purpose/intention it is collected.

Overuse of Data

The overuse of data can numb the individual intellect and diminish the person's decision-making capacity. Studies say that the leaders who engross themselves in data lose their potential to understand how their decision will affect the real world.

The use of data is a blessing for better decision-making, organizational growth, and customer satisfaction. Sometimes overuse of data leads to data-driven disasters. Let us take an example of a modern aircraft with numerous technological features that allow the system to make decisions based on real-time data. Now imagine an aircraft with a pilot equipped with all the AI Space expertise; the pilot is dependent on the system for decision making. If the flight equipped with state-of-the-art AI features experiences a major technological glitch, everyone will expect a well-trained pilot and a cabin crew. Even if given an offer to fly on modern aircraft without a cabin crew and pilot on half the price of the ticket, would you be interested in buying it? The human in the loop in this situation is essential for decision making. The ability to understand and make decisions is the paramount quality of human beings. Whether an aircraft or a car, machine-assisted and automated devices may sometimes become the data disaster victim, so it is a matter of concern in a data-driven society. A balance of data-driven decision-making and human decision-making is required to run the system.

1.9 Data and Information Technology Act

Data means presenting facts, figures, information, concepts, or instructions prepared and processed to get insight to help in the decision-making process in any form such as printouts, magnetic or optical storage media.

Under Section 65B(1), notwithstanding anything contained in the Evidence Act, any information present in the form of printed paper, stored in the electronic record produced by a computer system is deemed to be a document if it satisfies the condition mentioned in Section 65B(2). If the conditions under Section 65B(2) are satisfied, the paper on which the information contained in an electronic record is printed or any electronic media produced shall be admissible in any proceedings, without proof or producing the original, as evidence of the contents.

The council of Europe defined in the 1335th meeting of ministers' deputies adopted by the committee of ministers on 30 January 2019 the "electronic evidence" as any evidence obtained from data contained in or created by any device. They came up with the following guidelines concerning electronic evidence. They stated that the court should not refuse electronic evidence and deny the legal effect. The value of metadata of the evidence can help the court identify the origin and history of the evidence along with the date and time. The evidence can be submitted electronically rather than a hard copy on paper. The electronic evidence should be collected securely and offered through reliable, trusted services. The devices used for transmitting the evidence should be capable of maintaining integrity. As far as the national legal system permits and subject to court discretion, electronic evidence should be accepted unless authenticity is challenged by one of the parties and the reliability of the electronic data is presumed if the signatory's identity is validated, the data integrity is secured.

The Egyptian Law No.175, 2018, about information technology crimes, defines electronic evidence as to any electronic information that has the strength or has a significant value stored, transmitted, or acquired from the electronic devices which can be collected and analyzed by special application, software, or hardware.

The 'Personal Data' means any information related to an identifiable natural person. An identifiable natural person can be identified in one of the following ways directly or indirectly, particularly by referencing an identifier such as a name, an identification number like Social Security Number, Driving Licence, Voter ID, and location information. In certain circumstances, someone's IP address, eye color, political opinion, and a job could be considered personal data.

1.10 Case Study

Cambridge Analytica and Facebook

The social media platform Facebook and data analytics company Cambridge Analytica (CA) was in the news because of not protecting the users' privacy. It has been reported that CA harvested data from millions of Facebook users without their consent. It included personal data like the location of the user, page reactions, posts, and comments, which helped build psychological profiles that were used to analyze the characteristics and personality of the user. The data was collected through an app named "This is your digital life." Cambridge Analytica then arranged an informed consent process of the academic research in which they collected data from millions of survey respondents from Facebook and their friends [6].

This harvested data was later used in the political campaigns of the president contestant. The team used the data to micro-target the user by displaying customized messages about Trump to different voters on different social networking platforms. The advertisement was categorized based on supporters and swing votes as the swing votes play a vital role in elections. The CA's key was identifying those who might be attracted to voting for their clients or discouraged from voting for opponents. The supporters of the client were shown winning visuals of the candidate and the information about the polling booth. The swing voters were shown the visuals of the outstanding supporter of the client and negative images and news about the opponent.

Cambridge Analytica has expertise in building psychographic profiles, which identify individual personalities, and the information was used to influence the target audience. Facebook and CA came to the limelight to use these techniques to influence American voters in the 2016 election. In this campaign, the client for the company was the president's party, and the company was successful in its targeted advertisements. The primary concern in the case study was data harvesting of Facebook users without their consent.

Reflective Questions

- How can we find that the data is ethically available for its intended purpose?
- Does the organization inform the users when the personal data are collected?
- Do we take appropriate care when sharing or collecting data from third parties?
- If the data about x is collected by y, who is the data owner?
- What could be the probable consequences of overuse of data? Explain.
- Explain Data Theft. How can we avoid it?
- How can we utilize environmental data to avoid disasters?
- How is Big Data helping society to grow in a positive direction?
- Explain agriculture analytics. How can GIS help the farmers to increase their productivity?
- Do you think Electronic Evidence should be allowed in court proceedings? Why or Why not?

References

1. A.P. King, P. Aljabar, A.P. King, P. Aljabar, *Chapter 5—Data Types* (2017)
2. T.M. Cover, J.A. Thomas, Elements of Information Theory. (2005). https://doi.org/10.1002/047 174882X
3. C.S.R. Prabhu, A.S. Chivukula, A. Mogadala, R. Ghosh, L.M. Jenila Livingston, *Big Data Analytics: Systems, Algorithms, Applications* (2019). https://doi.org/10.1007/978-981-15-0094-7
4. L. Igual, S. Seguí, *Introduction to Data Science: A Python Approach to Concepts, Techniques, and Applications* (2017)
5. M. Kapil Tiwari, "icnl-19q3-p2-v1," *IEEE India Info* **14**(3)
6. M. Boler, E. Davis, *Affective Politics of Digital Media* (2020).https://doi.org/10.4324/978100 3052272

Chapter 2
Data Privacy

*The right of personal privacy is precious. Without it, we are
potential victims for a prying secret police.—Lewis B. Smedes*

Privacy, in the essence of the word, is a personal entity, which when breached can
have implications for the individual and society at large. Violation of privacy indicates
that something is amiss in a broader system that is used by or affects several people.
Its wider implications have enabled it to be recognized as a fundamental right by the
United Nations (UN) and other global and regional treaties. Privacy is the individual's
right to be safeguarded against intrusion into their personal life by disclosure of data
or by direct physical means. In an era where information is transferred to third party
applications actively, claiming privacy of data has become essential to assert when
and how much of one's data is sent to others.

Privacy is valuable in today's context as it is an essential component of safety,
in both physical and economic contexts. When civilians feel safe, they respect the
service provider, invest more in their services and feel free to communicate and live
without active self-censorship. Privacy has indirect financial value in the sense that
as an entity, it protects critical data, which when breached can led to large scale
monetary loss. Some people may also choose to pay more for a service that has
better privacy features.

The underlying risk of privacy breach makes it is essential to consider its monetary
and safety value as integral to the proper functioning of an organisation or society as
a whole. Although privacy has value, it can also be exploited to hide misinformation.
Crime and social injustice thrive on the discreteness of privacy, so its value is then
determined based on the ethical nature of its use. It doesn't have value innately, its
value exists in relation to external or personal motivations. The value of privacy is
derived or exploited depending on its use.

Consumers trust companies that keep their data safe and private. They see privacy
as being one of the core values of the company, especially in financial service sectors
or the healthcare industry. In a study on consumer data and privacy, it was found
that the media and entertainment sector do not earn substantial amounts of consumer

S. Shukla et al., *Data Ethics and Challenges*, SpringerBriefs
in Computational Intelligence, https://doi.org/10.1007/978-981-19-0752-4_2

trust [1]. This could be because privacy isn't at the foundation of their function-ality. Their marketing message doesn't revolve around securing the trust of the individual. Whereas for commercial businesses, customer loyalty and trust is linked to their income, which is why it is essential that they show the audience the narrative of privacy as their core value. Such organisations that deal directly with the customers are more likely to be under the radar for privacy breach or misuse. There is more surveillance by the media on these institutions which may bring out news of violations. Such news may give out the message that they don't value privacy or that the consumer lacks trust in the organisation. Understanding how privacy works for organisations and the individual will help in viewing its different aspects and implications.

The structure of this chapter is as follows. In Sect. 2.1, different aspects of data privacy such as its scope, principles, and challenges are presented. Definitions and differences between concepts are also discussed in this part of the chapter. In Sect. 2.2, various perspectives and approaches to privacy are discussed. The laws and legal regimes of privacy are presented in Sect. 2.3. Modern principles emerging in current legal practice of data privacy are also included in this section. In the final section, the models of privacy are presented with two case studies followed by a critique of the existing legal regimes.

2.1 Data Privacy: Aspects, Challenges, and Definitions

The relationship between data gathering and transfer, tech systems, the public's expectation of privacy, and the political and legal concerns surrounding them is known as information privacy. It's also referred to as data privacy or data protection.

Understanding Data Privacy

Data privacy dictates how data is shared, stored, and used across different channels. It is a type of privacy of personal information that is offered to private actors in a number of situations and contexts. It is a process in which ones information is transferred, stored, or assessed. While preservation of personal data is an important component, the use and potential misuse of the data is also considered. Understanding privacy at its basics will assist in learning about its various applications.

Meaning and Scope of Data Privacy

Data privacy means that there are ethical standards when collecting, processing, sharing, archiving, or deleting data. It also covers individual rights, the purpose of data collection, privacy choices, and how organisations manage personal data of subjects.

When we discuss the scope of data privacy, we must look at the benefits and its potential impact on information systems. As stated in Fig. 2.1, the following are different aspects that depict the scope of data privacy and its wide-ranging possibilities.

Fig. 2.1 Scope of data privacy

- Earn Employee Trust

The scope of data privacy is evident in brands and companies using it to gain employee trust and to facilitate consumerism. The ethical use of data will become an attractive feature for customers and will lead to gaining trust on a large scale level.

- Legal Responsibility

Data privacy also keeps companies aligned with certain legal responsibilities and ensures that they abide by widely accepted and lawful modes of operation of data privacy.

- Prevent Hacking

Data privacy can prevent excess information about an institution from reaching third parties. This reduces the chance of getting hacked and thereby hinders people from getting access to assets of an organisation.

- Secure Assets

Secure data transfer will also lead to better user experience in case of streaming services like Netflix and other data-driven applications. Since privacy has come under the radar in recent years, companies have begun taking actions actively towards preserving data privacy, which means that more secure transaction is underway and will impact several sectors like healthcare, ecommerce and others that deal directly with consumers.

Principles and Laws that Govern Data Privacy

There are principles that are integral to data privacy and should be considered essential in matters that concern the use of personal information. The principles include notice or awareness, choice or consent, access or participation, integrity or security, and enforcement or redress [2]. These were put forth by the Federal Trade Commission for privacy protection.

- Notice or Awareness

Systems and websites should inform users about ownership, security, and terms of service. A banner that appears at network log-in, advising that access to the network is restricted to authorised users, is one example of such a notice. It could be a splash screen for a website that informs users that entering indicates acceptance of the terms of service.

- Choice or Consent

Giving customers the choice over how their personal information is handled. This is in reference to secondary uses of information, which the FTC defines as uses other than those required to consummate the anticipated transaction. Such secondary uses can be hidden, such as adding the customer to the collecting company's mailing list in order to promote more products or promotions, or external, such as the transmission of information to outsiders.

- Access or Participation

Users should be allowed to learn more about how their data is stored and to challenge its accuracy and completeness if they believe it is incorrect. Data collectors must take reasonable precautions to ensure data integrity, such as using only reliable data sources and cross-referencing data against various sources and enabling consumer access to data.

- Integrity or Security

The accuracy and security of data is the fourth principle. Collectors must take reasonable precautions to ensure data integrity, such as using only reliable data sources allowing consumers access to data, and discarding or converting untimely data to anonymous form. Security includes both management and technical safeguards against data loss as well as unauthorised access, deletion, use, or disclosure.

- Enforcement or Redress

Enforcement refers to the fact that particular privacy rules will apply to your website. Your company may join an industry code of practice or a privacy seal programme, both of which may contain dispute resolution methods and penalties for noncompliance with the program's criteria.

Existing Laws on Data Privacy
As technology advancements have enhanced data collection and surveillance capabilities, government bodies have begun to pass rules on what data can be gathered on users, the different uses of that data, and the secure storage of the information. The following are some of the most essential regulatory privacy regimes to be aware of:

- Sector-Specific Privacy Laws: While there may be no general data protection legislation in the United States, there exists laws and acts that protect the information of certain sectors. Child information is safeguarded at the federation level by the Children's Online Privacy Protection Act. The Health Insurance Portability and Accountability Act (HIPAA) aims to ensure that personal healthcare data is protected and handled with care. There are acts that also protect online-streaming and audio and visual data.
- General Data Protection Regulation (GDPR): GDPR is a regulation that mandates enterprises to protect European Union (EU) citizens' personal information and data when conducting business within the EU. Noncompliance could cost businesses a lot of money. Basic identifying information such as address, name, and phone numbers, online information IP address, cookie data, and genetic data, biometric information and political beliefs are all protected by this law.
- State Laws: The state laws impose regulations on businesses to control the way they collect, disclose, or transfer data like medical records, email IDs, and other financial information. They can also protect public sector data like tax information and other sensitive data.

Challenges Faced in Online Privacy Protection
While there are laws that govern data privacy, the online availability of data has put sensitive information at high risk. As technology can have gaps and some information is publically known to be of high value, data is facing several threats due to the following challenges:

- Site Tracking

Online algorithms have led to constant tracking of consumer behaviour. Ecommerce websites and blogs have backend systems that monitor site action, which many users may not be fully aware of. Data may therefore not be protected to a degree that users may be comfortable with.

- Data Transfer

When terms and conditions are accepted on websites, there may be certain clauses that may let data be shared beyond the site that users interact with. Websites have partnerships with certain applications that may use the data collected from the site. Some websites may also have dense and intricate privacy policies on data transfer that are exhaustive and complex to understand.

- Threats due to Cybercrime

Many attackers attempt to steal user data in order to conduct fraud, infiltrate security systems, or sell it to parties who will use it for nefarious purposes. Phishing attacks are used by some attackers to deceive people into disclosing personal details via a false email or fake advertisements on websites.

- Misuse of Social Media

Currently, it's simpler than ever to track down anyone on social media, and users' comments may expose more private details than they know. Furthermore, social media networks frequently collect more data than users realise. Photo-based posts can reveal more personal data unknowingly like location and lifestyle habits, which makes privacy a personal principle and responsibility too.

Privacy and Society
The implications of lack of privacy or over-infringement has indirect effects on society as a whole. For state governments, privacy can enable safety against national threats like terrorism. The private sector may benefit from breach of a customer's privacy and use the data to create better marketing strategies, which will eventually impact the economy. Personal privacy and its breach can have positive and negative impacts for the society.

China's Mass Surveillance
The enormous network of systems employed by the Chinese government to monitor the civilians of China is referred to as mass surveillance. While being monitored constantly when in public can help in matters of preventing crime and justice in legal trials, it also means that someone is taking note of one's movements publically 24/7. Surveillance may not be evident and may violate an individual's notion of personal privacy.

In 2014, China's governing cabinet also known as the State Council sent a public notice for the setup of a tracking system across the nation to assess or rate the reputations of people, commercial businesses and government officials. China's Social Credit System also forms part of the surveillance. China's Zhima Credit is tied to Alipay, a mobile payment platform. The app is used to pay for anything from a supermarket bill to a restaurant cheque. As convenient as this may seem, the underlying fact is that every purchase along with location is tracked.

The data generation done by social credit can hinder people from purchasing things beyond their affordability. It can also indirectly stop social events and bring a halt to exchange of communication between the user and certain webpages. When privacy and society are discussed, consent forms a major part of the ethics in the process. Downloading and approving certain functionalities within an app may grant access to further privacy breach. When consent isn't taken, it will violate customer trust and the foundation to build healthy customer-operator relationships.

Society is also shaped by the cultural views of privacy. Both concepts impact each other. Some cultures believe that individuals have the liberty to live, speak, dress and act as the please, lawfully. While other cultures may assume that they have the liberty to interfere in one's business or spatial boundaries. In case of the latter, distrust and miscommunication can arise, while in the former, a community that is more accepting and progressive can form which will also benefit the economy. The collective view of privacy impacts society.

Protecting Privacy in Information Systems
Information systems are a set of components that gather and share data. Protection of these systems are enabled through clauses mentioned in laws and regulations. Having a framework for all companies to follow will streamline the data protection laws, however, as information systems are complex and have several layers, multifaceted data laws are required. To safeguard data, it is essential to classify it in different levels as requiring different operations. The following methods are suggested to protect privacy in information systems:

- Safe Website Browsing

Anything you do online can be tracked by the big data-driven companies. Different online tools can collect data on location, interests, and other factors. By blocking advertisements and the data they collect, browser extensions can help secure information systems. A Virtual Private Network (VPN) can benefit users who use open sources of internet since it gives an extra degree of security to web.

- Password Managers

A trusted password manager can enable privacy by suggesting unique and complex passwords for each account. Some users may use easy passwords so that they can remember them when they log into new systems. However, this poses a threat and makes accounts vulnerable to hackers. With a password manager, security breaches can be monitored and weak passwords can be changed.

- Antivirus

Viruses aren't as frequent as they were a decade ago, but they're still around. Malicious software on computers can cause all sorts of problems, from intrusive ads to personal data scanning. It's important to install an antivirus software when a computer system is shared with numerous people in the organisation or there's a risk of opening dangerous links.

2.2   Privacy and Confidentiality

When it comes to technological tools, the terms privacy, confidentiality, and security have a lot of similarities, but they also have distinct meanings and roles to play when it comes to data upkeep and administration.

Definition of Security, Privacy and Confidentiality
Security is defined by a state of being free from threat or danger. Privacy is the right to keep certain information about yourself private, while confidentiality is the right to keep personal information or sensitive organisational data from being disclosed to others.

Objectives of Security, Privacy and Confidentiality
Privacy allows people to set limits on who has access to a person's personal and private information. The goal of confidentiality is to prevent unauthorised access and misuse of information. The goals of privacy and security are sometimes the same. In some circumstances, security may not be sufficient to address privacy issues. A corporation or government body may be prepared to hold its data secure from external attackers, but workers may be able to view customer data.

Difference Between Security, Privacy and Confidentiality
The term "privacy" refers to a consumer's right to keep his or her personal information from been seen or used by the public. It entails preventing sensitive data from being openly shared or sold to other parties over the internet.

Confidentiality, on the other hand, is used when people refer to someone who has been entrusted with personal information that must be kept private. Alternatively, some may describe confidentiality as concerns about the data obtained, whereas privacy matters are concerned with the fundamental idea of an individual not being recorded or observed.

Enterprise or government systems use a distinct phrase called security. The overall purpose of security is to secure an organisation or agency that may or may not have a lot of sensitive consumer or client data.

Security, Privacy and Confidentiality: Comparison Chart
All the three entities deal with securing information from invaders and unlawful access to information, but, as stated in Table 2.1, the difference lies in the type of information that is protected and the measures set in place to ensure that data is safeguarded.

The assets of a company may have robust security, but when documents are shared between colleagues, confidentiality is often the responsibility of the employee. Security can be set up by a software or application, but confidentiality deals with the private sharing of information despite a software system. Privacy on the other hand is an ongoing process. Extraction of data and disclosure of it is governed by cookies and terms of conditions and consent of the user is a critical point in its functioning.

Table 2.1 Difference between security, privacy and confidentiality

Security	Privacy	Confidentiality
A system that safeguards information	The state of being free from public intervention	Owning access to shared but private information
Provides protection for different types of information and data	Protects sensitive information related to organisations and individuals	Prevents specific data from being shared with outsiders
It sets certain protocols in place to secure assets	It deals with protection of certain rights of privacy with regards to personal information	It details norms to protect sensitive documents and assets of an organisation
Focuses on all kinds of information of an organisation or system	Focuses on processing of owned information	Focuses on keeping certain information between select employees of an organisation

2.3 Privacy as a Philosophy

As privacy is asserted as a basic right, it is a given that it is also a philosophy that has been debated over as a concept for decades. Privacy dictates how individuals communicate and interact with one another and also encompasses various ideas of the self, control of information and disclosure which are integral to life. The idea of privacy significantly changed over time and the change was particularly evident in the last decade, especially among companies and individuals with broad access to communication technologies.

Privacy—Historical and Cultural Views
A cultural purview of the idea of privacy across time will show how the concept has evolved in the individual's mind to assume the aspect of risk. Earlier when risk was low, the idea of privacy was less developed and was considered covered under general human rights. Now, with the advent of social communications and information networks of varying complexity, privacy is controlled and operated beyond the reach of the individual.

Privacy Through the Years
In pre-modern times, when human interactions were face to face, the idea of breach of privacy was dependent on one's own terms by setting personal boundaries and sharing generic information. Both the individual and the potential offender were present at the same time. With the advancement in technology, data can be collected, stored and searched in large quantities at any period of time during or after the interaction, which is where distrust or lack of transparency affects the individual's idea of privacy.

In 1967, legal scholar Alan Westin discussed reductionist accounts that held the importance of privacy to be related to external values and sources of value. The opposing view maintains that privacy is valuable in itself and this statement can be

Fig. 2.2 Philosophical questions on privacy

addressed through questions in Fig. 2.2. In this alternative approach, the value and importance of privacy are not derived from other considerations. Views that construe privacy and the personal sphere of life as a human right would be an example of this non-reductionist conception.

Towards the latter half of the twentieth century, newer debates arose regarding privacy as a human right. The questions posed were around its emergence and its justification philosophically. 'With which entity does the responsibility of protecting privacy lie?' was also discussed. This notion of generic privacy is normative, and some scholars argue that it may conflict with other cultures' social and legal systems, and thus cannot be considered absolutely. When one studies the origins of universal privacy as a right in legal and political conceptions, the evolution of this dispute is clear.

Towards the end of the twentieth century, themes dealing with privacy began surfacing in court cases. Smith, H. Jeff, et al. mentioned in their research study on information privacy that the common themes included the idea of the self and general privacy with respect to abortions and embryos, privacy and its relation with the press in respect to exposure of private details and intrusion in the private lives of celebrities, and voyeurism and monitoring of lifestyle in workplaces [3]. Court cases with such themes raised two significant issues, the need for a more specific definition of general privacy than the "right to be left alone," and the state as the guardian of general privacy. Because of these two factors, the general privacy argument among legal and political specialists became ideological.

Culture and Privacy

The culture of a society determines the mode in which data is collected and transferred and the intention is determined in this manner too. In rural parts a country, for example, a lot of information dissemination happens orally via radio transistors or television, so the concept of data privacy has little scope there. For cities that rely on applications to share sensitive information, data privacy becomes important. This also has to do with the income of the individual in a metropolitan area, their spending behaviours and tech savvy nature. The culture of a place therefore can induce or eliminate the need for data privacy. The idea of culture can also be in one's household or workplace can also influence if privacy is maintained or breached. Toxic workplaces may induce distrust and lead to invasion of privacy, while a healthy work environment can foster respect for boundaries and personal information.

In the United States, until the twentieth century, common law did not recognize any right to general privacy. The U.S. being a tech-first country would have several limitations with a strict privacy law in place. It could have possibly hindered the growth of several social media applications that are now openly data driven. The culture of a space impacts the privacy laws which also explains the diverse interactions with privacy around the world.

Critiques of Privacy

American philosopher Judith Jarvis Thomson observes privacy as a mere combination of different rights. Property rights and the rights to bodily security include the rights of privacy. She asserts that privacy is derivative and the breach of it is simply a violation of a basic right. Richard Posner provides a critique of privacy from an economic perspective. He states that selective sharing of information can induce concealed economic gain. The individual may exploit privacy rights for personal monetary advantages which won't enhance the economy. Posner states that privacy is essential for corporate organizations as they contribute to the economy, while personal privacy can affect the larger monetary equation [4]. A feminist critique of privacy also exists which looks at privacy as being a disadvantage for women when it is employed to hide information about unequal pay in the workplace. It is also looked at as being harmful when there's concealment of abuse. This criticism looks at privacy as actionable and manipulative.

Threats to Privacy in the New Technological Regime

Privacy law has evolved in sync with technology, reshaping itself to address the privacy dangers posed by emerging technologies. The information revolution, on the other hand, has occurred at such a rapid pace and has affected so many aspects of privacy legislation that the traditional, adaptive legislative and judicial process has failed to appropriately address digital privacy issues.

- Behaviour Tracking and Liberalization of Data Market

The new technological regime revolves around applications and systems that rely on personal information of their users to provide a profitable customer experience. This has given rise to threats like increased data sourcing and the subsequent collection of enormous quantity of personal sensitive information. Every monetary transaction or online communication can now be recorded and stored. Another threat is the liberalization of the data market and the possibility of accessing and using personal data. With new products emerging rapidly, there's a lack in varieties of control mechanism built to fit existing and upcoming technologies.

- Misuse of Data Transparency

While transparency may gain customer loyalty, it can also be an entryway for hackers. When systems reveal how they protect data, for instance, by using masking, this exact information can be used to infiltrate the system and access data. Hackers search for vulnerabilities or loopholes in the masking method. Such kind of access to data is called a transparency attack.

Approaches to Privacy
Since privacy affects a large set of people and at different levels, there are several approaches to the concept. Each of these ways of understanding privacy are interrelated and have varied outcomes.

- Individual Approach

The individual approach to privacy involves self-management. The individual takes responsibility to check terms and conditions, examine website security and take control over their personal data. This approach involves several rights like the right to removal of personal data and procuring information about how data is being stored.

- Social Approach

Another approach to privacy is through the social lens when we see the seriousness of the breach of privacy on the society and look at how it can be addressed. Priscilla M. Regan in a study, makes a point that politicians and the society at large agree that threats to privacy have occurred after the increase in organization-individual relationships [5]. The approach moves from the individual realm of breach of personal knowledge spaces to highly sensitive information that if leaked, can impact wider groups of people via an institution. Organisational problems have led to thinking of privacy from the social approach and looking at ways to prevent major threats.

- National Approach

Privacy is also studied in its effect on the nation. Safety of a country is at the core of the national approach to privacy. Keeping data of the nation private and secured from any kind of infringement, whether foreign or within the country, also helps with real time security and protection at the borders. Privacy also gives civilians the freedom to share thoughts in communities, which indirectly contributes to unity. When people can express themselves and find other people with similar ideologies in a safe and accepting manner, the sense of belonging is established. While nationhood may seem like a narrative driven by coercion, it is freedom to think and keep thoughts and ideas of cultures protected within communities that brings a nation together.

2.4 Data Privacy Law in the United States, European Union and India

Given the reliance of companies on data, provisions have been made in different countries to enable privacy for users and organizations. Countries usually employ a set of different privacy laws to deal with various aspects of data concerns. In the US, the Gramm-Leach-Bliley Act (GLBA) protects citizens against violations in the financial sector, while the Children's Online Privacy Protection Act (COPPA) secures the personal data of minors. These and other acts come together to form the privacy regime in the US. In contrast, the European Union has enabled a more centralised law, the General Data Protection Regulation (GDPR) to regulate the personal data of its citizens. It includes consent for the processing of information of the citizens and secure cross-border transfers. In India, currently, the Information Technology Act is an important decree on data privacy judgement. It maintains that companies that work with sensitive personal data should have security measures related to such information in place.

While differences exist across data privacy law across the world, there have certain high-focus areas in common. In emerging law, definitions of privacy-related terms have been given significant focus. Personal information, for example, is detailed as data which can aid in identifying people. Sensitive personal data, more specifically, is noted as a subcategory that branches out into passwords, financial data, health, sexual orientation and biometric information. The difference between the two has been integrated in the legal framework of different countries and this is one way where privacy can become more robust and highly effective.

Modern Principles of Privacy Law
As per the existing regimes on data privacy, certain measures aim to create a culture of protection of the individual's liberty and to ensure that personal data is used securely. These can be identified as principles of privacy law in the modern era which are advanced and suited to the current technological landscape. Following are the principles that can be identified across countries.

- Categorizing of Data

Accurate categorization of data and provisions for types of data is an emerging principle in data privacy law. Personal data refers to an individual's name, location, religion, and other individualistic details and attributes. Non-personal data refers to information that does not have any personally identifiable particulars like land records, municipal data sets, and vehicle papers. The management and handling of this information has been given more focus as its application is broad and aids the functioning of the economy. Categorizing such data enables individuals to understand what kind of information is being shared and what is protected.

- National Security cannot be a Free Pass

The unethical breach of privacy of citizens by the government under the name of national security is not considered a reasonable action by the legislative bodies. When the fundamental right to privacy is breached, the government has to the legal authority.

- Focus on the Individual

The law on privacy currently focuses on the right of the citizen to choose who can keep watch and when. The aim is to give back to the individual the right to liberty of life. In the U.S., the Federal Trade Commission Act safeguards consumers against theft and other unfair practices. Strict compliance to privacy terms and appropriate management of data are in place. This is also how the US government was able to press huge fines on Uber and Facebook.

- Transparency in Data Processing

Data transparency brings clarity and straightforwardness in how different organizations deal with data. As a principle, it makes it necessary for companies to be clear on how they collect and process sensitive personal data. Before signing a contract on employment, the company must make evident the information that will be transferred to any partners of the business.

- Experiments through Sandboxing

The term "sandbox" refers to a controlled setting where innovative products can be tested. In the context of data protection, a sandbox allows a company to collaborate with the DPA to test new products and services for compliance with data protection rules. The experimenting principle allows companies to improve their policies and user experience without risking liabilities. Sandboxes are also gaining popularity among European DPAs. In 2020, the Norwegian DPA set up a sandbox initiative to aid the growth of emerging AI technologies that had ethical foundations.

- Localisation of Data

Although data localisation can refer to any prohibitions on cross-border data transfer, it has largely come to refer to the need to physically locate data within the country. The PDP Bill allows personal data to be transferred outside of India, with the exception of sensitive personal data, which must be duplicated in the country (i.e. a copy of the concerned data will have to be kept in India). Data processing entities, on the other hand, will be prohibited from transmitting critical personal data beyond India.

Legal Regimes for Protecting Privacy
As discussed before, the privacy law varies between countries and this difference has to do with their technological development. With increasing focus on innovation, there are relaxed laws in some countries, while others have secured systems with a robust and all-encompassing law in place because of the magnanimity of loss a

breach could cause. Let's look at these different laws to understand the aspects they entail.

Data Privacy Regime in the US

The privacy regulations in the US cover the health sector, financial sector, children's data, and some state-level laws. In this section, two state-level regulations have been discussed to understand how the states look at data processing on the internet.

- California Consumer Privacy Act (CCPA)

The CCPA safeguards online activity as it details privacy protections for users of the internet. This act grants users the right to ask businesses about the personal information that is being stored by them. The act mandates that businesses can't transfer their information to third parties without consent and users have the right to access their information, correct it or delete it.

- Nevada Senate Bill 220

This bill is also directly aimed at internet privacy. It mandates that websites or e-services are prohibited to sell data about customers in the State without the affirmation or knowledge of the consumer. The Nevada law used the term "covered information", which in this case includes name, address, email id, contact information and social security number. The sale of covered information without consent is prohibited and consumers can choose to opt out from such activities.

Data Privacy Regime in the European Union

The European Union currently regulates personal data with a robust law in place and is a model that is desirable for consumers as it gives users rights and protects them against various threats on the internet.

- **General Data Protection Regulation (GDPR)**

The GDPR has the mainstay on sharing personal data in and beyond the European Union. The main impact the law has is that individuals from the EU gain control over how their personal information is handled across the web. The GDPR has made it difficult for businesses in Silicon Valley with EU consumers as they have to comply with strict data processing and change their operations.

- **The European Data Protection Board (EDPB)**

The EDPB consists of the head or representative of each DPA and also the European Data Protection Supervisor (EDPS). The role of this group of individuals is to ascertain that the law on data protection is followed through across the EU. They also send out guidelines on the core ideas of GDPR and play a part in ensuring that there is consistent application of the rules across different jurisdictions.

Data Privacy Regime in India

In India, if the Information Technology Act alone doesn't help a case, the rights of the citizen detailed in the Constitution of India take centre stage. Currently, the Personal

Data Protection (PDP) bill is proposed to introduce limits on data collection, content-related processing, rules on third party transactions, and handling of personal data. Let's look at what India has in place currently:

- **Article 32**

A right without remedy can make a fundamental right less effective for the individual, but Article 32 is very carefully drafted. Under sub clause is (i) the right to move the Supreme Court by appropriate proceedings for the enforcement of the rights conferred by this Part is guaranteed. Nirmalendu Bikash Rakshit explains it as 'an aggrieved person can move the Supreme Court for a legal remedy in case of an alleged infringement of his Fundamental Right. In such case, it is the constitutional duty of the apex court to look into the matter and to provide necessary redress to the affected person' [6].

- **The Information Technology Act**

This act sets mandatory regulations in place for body corporates like firms, companies, sole proprietorships and other communities of professional relations that deal with Sensitive Personal Data or Information (SPDI). The SPDI rules mentioned under the IT Act, indicate the standards of protection required for such kind of data. They maintain that companies should have a privacy policy, get the consent of employees while collecting such data from them, and to notify data subjects of reception of the data that is gathered. The act also asserts that body corporates are liable to pay for damages or distress caused by lack of security measures from their end.

- **Article 21 and Right to Privacy**

Article 21 mandates that 'no person shall be deprived of his life or personal liberty without due process of the law'. This statement can be extended to include the right to privacy with regards to property, health records, and personal information. No body or organisation is permitted to access such information unlawfully.

Privacy as a Legal Right

The legality of privacy can be discussed by looking at what the law mandates as the right procedure or behaviour in terms of privacy. In several countries, as a legal right, a person has the freedom to autonomy, limited and protected communication, and minimal exposure. The person also has the right to move freely, with freedom of speech and expression. The word freedom here implies that there should be no unlawful restriction in this freedom and breach of privacy violates freedom. Consent forms a major part of the legality of privacy, as it brings in transparency and accountability, which are principles of the data privacy regulations in India.

Privacy is also protected with criminal liabilities. Intrusion into one's personal property has criminal charges, which indirectly also means that personal privacy is secured under the law. This brings us back to the critiques of privacy and the perspective that privacy is a collection of different rights. Is it really necessary to have a law on privacy? The thought can be countered by another: if there is no body on privacy regulation and management in India or no concrete law, the idea will be stagnant, the definitions eventually will be outdated, and it may not progress or be relevant to the complexity in the advancement in technology.

2.5 Operations and Concerns of Privacy

Having an approach to privacy determines how sensitive and critical information is collected and distributed. The models of privacy present different levels of operation. From conservation to collection and behaviour mapping of data, these models are also points of concern as they conceal power structures, surveillance, active tracking and authority.

2.6 Models of Privacy

Looking at privacy as a model or perspective enables users to understand what is being captured. Depending on the type of model, the data that will be stored and transferred will be different. Such frameworks help users assess how privacy works and how their information is used. The following three models of privacy have largely been noted by Jens-Erik Mai, a Professor of Information and they cover three different perspectives [7].

The Capture Model
The capture model is based on the idea of the structure of the action. In this model, the focus is on how human activities are converted in representation languages of computer systems. It employs structural metaphors and describes the captured activity as derived from a catalogue of parts. The capture model is distributed and varied, with activities taking place inside of particular local practises.

The Panopticon Model
The panopticon model is a metaphor for keeping an eye on things. This is the standard view of privacy and surveillance. The core premise is that surveillance and invasions of privacy are carried out by someone "watching" another person. In this model, the assumption is that the observing is nondisruptive and covert. According to Foucault the panopticon's major impact on the inmate is to promote an awareness of permanent visibility that enables the automatic influence of power.

The Datafication Model
The datafication model employs behavioural tendencies. While both the panopticon and the capture models of privacy place a priority on data gathering, the datafication model puts a focus on data processing and analysis. The datafication model is concerned with the facts or realities that data can provide after it has been processed and analysed.

2.7   Information Loss

Information loss is a privacy concern because when data is lost one's liberty to choose to anonymous or to preserve their data is also lost. Information loss happens when valuable data is deleted or removed due to an error either by malware, equipment, or by employees at an organisation. Human error related data loss may happen intentionally within an organisation. For example, a sales manager of the company may manipulate information in the data sheet if the salesforce under him or her is not meeting their monthly sales targets. They may forge or delete sales information which may cost the company in the long run.

Data loss may also occur unintentionally. The UK Prison System, for instance, faced a major information loss when an employee lost a USB drive that contained sensitive data including the release dates of over 84,000 prisoners. As a result, their reputation was affected and they lost a major client, which also indicates how responsibility and physical security are an important part of keeping data safe.

In 2007, the company DreamHost lost over five hundred websites because of a bug that had infiltrated their routers. Although they addressed the issue, their records were still being removed by the bug. The company had to apologize to their customers. However, if their system had been built with a strong security system, they wouldn't have lost their files due to malware. As seen in Fig. 2.3, several factors lead to data

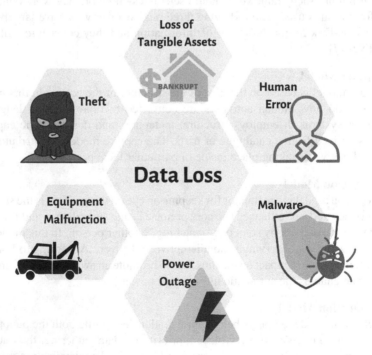

Fig. 2.3 Causes of information loss

loss and these should be considered when opting for a security plan or a privacy protection system. Power outages, equipment malfunction and theft can be avoided by having backup and preventive measures in place.

2.8   Infringement of Privacy

Applications and software that have their databases online are exposed to the threat of theft. Infringement is one of the main violations of privacy and it is an operational concern as it occurs due to unprotected data. Take the Coinbase Hacking Incident of 2021 for instance. In October 2021, third party hackers found an error in Coinbase's SMS account recovery process and got into their account database. They transferred assets into crypto wallets that weren't affiliated with the company. This breach of privacy was on two levels, first the company's system and then the 6000 customers from whose wallets the money was stolen. Although they reimbursed the funds, the trust that their users place in the company may have taken a hit. The infringement of privacy leads to monetary loss and when made public, loss in the collective trust in the company.

A privacy breach may not always be related to data. Privacy may be violated even when an individual's personal information isn't respected. Sharing information to external sources that is said in confidence is also an invasion of privacy. For example, a manager might tell his associate about how he couldn't perform a certain task to his best ability because of personal distress. If the associate shares this private information to other employees, it is a breach of privacy.

By avoiding oversharing and keeping the device containing sensitive data secure with a security service provider can enable protection against infringement of privacy. If any data breach occurs, one should inform officials who can take immediate action and freeze your credit card and other financial assets to secure them from ill action. As mentioned in Fig. 2.4, using strong passwords with mixed characters will protect accounts from novice infiltration. Browsing safe websites can also help in avoiding theft of information. One's data can also be secured from breach by actively being aware of the information that one shares. Lastly, security is an important part of privacy and having a reliable software to protect data is also vital in safeguarding it from theft.

2.9   Case Studies

The selected cases are from two different countries so that the difference between privacy regimes is evident. This reveals loopholes and weak points in the policies and offers points for improvement. Here are two case studies where the outcome for a similar act of breach was different:

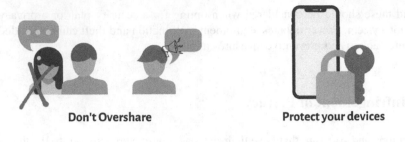

Fig. 2.4 How to avoid privacy infringement

- **Air India Data Breach**

In May 2021, Air India said that its customer database was breached due to a cyber-attack. Its IT infrastructure managed by SITA, a Switzerland-based tech company had faced the attack and around 4.5 million records were possibly leaked. They informed the affected clientele that the breach included personal data that was registered during August 2011 and February 2021, which is well over nine years of data. They also said that the breach did not affect password data. Information like name, contact, ticket, passport, credit card data, date of birth, and frequent flyer data was violated. Leakage of such a proportion can pose a threat to the safety of individuals involved and impact larger communities including journalists, activists, and well-known personalities with a social standing.

As part of its answer to the huge breach of security, Air India announced that it had investigated the security breach, secured the servers that had been compromised, worked with external data security incident professionals, notified and worked with credit card issuers, and reset credentials for its Frequent Flyer programme. They also said that their data processor detected no suspicious activity after the affected servers were secured, however this still hasn't been confirmed officially. While some passengers sued the airlines for violating their right to informational autonomy, the company hasn't reportedly paid a large fine for the magnitude of the breach.

- **British Airways £20 Million Fine on Data Breach**

In another instance, the British Airways was also fined for a major negligence on their end in protecting their customer's information and failing to meet the regulations of the data protection law. The Information Commissioner's Office issued a £20 m fine for failure in safeguarding financial and personal data of over 400,000 passengers. The investigation was done by the Information Commissioner's Office (ICO) and the fine was issued by them as well. This came in lieu of the 2018 cyber-attack where the airway's systems were breached and then modified to extract information as they were fed into the system. Two months had passed until a security researcher found that their system was violated.

After the investigation, it became evident that the airways could have prevented the attack easily given the measures present on the Microsoft operative system that the company was using. The attacker reportedly stole names, payment card details, CVV, and addresses. For its response, the British Airways said that the informed affected customers as soon as they knew and also paid claimants an undisclosed amount.

Comparison of Case Studies

When we compare both case studies, we realise three significant differences despite the similarity in cases. In the British Airways case, there's a role played by three entities in the outcome of the issued fine: the data protection law, the investigation by the Information Commissioner's Office, and the security researcher. In case of Air India, no fine was issued by a specific authoritative body and there is also no actionable and concrete data protection law in place. The fact that Air India wasn't liable financially for the data breach indicates that the privacy law in India is not developed to a state where it takes into account the severity and repercussions of such a massive security breach. Air India's customer base included journalists and other socially active individuals and breach of their data can also be a threat to their lives. In all fairness to the customers, Air India should have been liable for the breach and paid claimants so that they can take measures on a personal front to safeguard their information.

Companies dealing with login credentials, addresses and payment details need to ensure that all current and up-to-date security measures are in place and a security officer is supervising or performing routine checks to detect any weak points in the system. The two case studies also show that data protection is a proactive process and requires knowledge about existing data privacy law or data protection regimes issued by governing authorities.

Critique of the Existing Privacy Regimes

While there are provisions for data protection in place, the implementation of various terms under the provisions is not addressed well, given the complexities of privacy breach. The GDPR for example, uses the term "personal information" which may involve easily available data that is distributed across several systems. It is difficult to keep track of such data from input to output. Integrating applications to perform such tracking in data systems will be expensive for new businesses. There also exists ambiguity regarding data-related terms because several companies store data that may be considered private and public records too. The CCPA addresses this problem by letting companies themselves assess if their resources can be "reasonably" associated with individuals, however, the doubt about what is reasonable and what is not still remains.

Developing countries with no central privacy law may see large scale exemptions granted to the government to process data under certain scenarios. The government may access medical records with the reason of setting up adequate infrastructure to be prepared for an epidemic. Non-personal information can also be assessed when demanded from a direct service provider after consulting the Data Protection Authority. This allows access to data which can be re-identified and traced to extract personal data that may be critical to some individuals.

The two questions we should ask are: Is the concept of privacy so complex that we cannot decide where to draw the line on data breach? Are companies data-driven to an extent that following strict privacy measures will affect their revenue and the economy subsequently? To answer these questions one can look at a framework for Non-Personal Data (NPD) that highlights that such data, in the context of strategic interests, can be used to re-identify a person. Definitions of terms needs to be considered more seriously and concretely to protect the privacy of all members. Even the term 'sandbox' is ambiguous and contains risk of breach in its application.

Another criticism is that the user's subjective understanding of personal and critical information may vary and the existing law doesn't take this subjectivity into account. Consent in the answer to this concern. Thinking about the user more is the way to go. Enactment of the law on privacy is rooted in how much information is granted to the individual. Hence, the legal regime on privacy may fail if the citizens of the country are not law-literate. Users are at times uninformed about the actions they can take when their privacy is breached. The media has made law accessible on several accounts by looking at sub clauses in legal rights during cases including data breach. Working with the media or other modes of communication to let the users understand their rights is also necessary for law enforcement.

2.10 Conclusion

The ability of data to generate value and contribute to building the economy makes it essential for governments to regulate all areas of information dissemination. With the advancement of AI and extreme focus on data analytics, the world is moving into a data-driven mode of operation. The scale of growth in data technology makes it is necessary for the judiciary to have clear-stated laws in place.

Law can be enforced fairly when operations are streamlined. The critique and different operations of privacy depict the complexity of the concept and the lack of a standard process. That's why specifying the meaning of privacy-related terminology in the privacy regulations is essential as it gives definition to a term that is subjective.

Such detailing keeps privacy grounded and also helps with the enforcement of the law in different contexts. With advancements in technology, the definition of privacy, sensitive data and public data will continue to evolve. Having a regulatory body of data privacy experts to map these changes and suggest regulations that will adapt to new developments is the way to move forward.

Reflective Questions

- Given the varied and subjective meaning of privacy, is it possible to have a standard privacy law that treats all infringement cases in fairness?
- Should the government, service provider, or the individual be responsible for effective data privacy of the user?
- Explain different perspectives on privacy and give your own critique on privacy as a concept.
- What challenges arise with the increase in applications and software that are extremely data-driven and have third-party partnerships?
- Explain three points of difference between privacy, security, and confidentiality.
- Can a developing country emulate the privacy laws of developed countries or does it need a different privacy law based on its culture and technological advancement?
- Elaborate on the scope of data privacy. What aspects can be covered with its application?
- What loopholes exists in the current privacy regimes of the world? Refer to case studies to substantiate your answer.
- How can information systems be protected from data theft and privacy breach?
- Do you think the right to privacy is covered under constitutional law? Explain why the right to privacy should or should not be a standalone right.

References

1. V. Anant, L. Donchak, J. Kaplan, H. Soller, *The Consumer-Data Opportunity and the Privacy Imperative.* McKinsey & Company. https://www.mckinsey.com/business-functions/risk-and-res ilience/our-insights/the-consumer-data-opportunity-and-the-privacy-imperative. Accessed Nov. 18, 2021
2. M.K. Landesberg, T.M. Levin, C.G. Curtin, O. Lev. "Privacy Online: A Report to Congress". Federal Trade Commission. June 1998. Accessed: Nov 20, 2021 [Online]. Available: https://www.ftc.gov/sites/default/files/documents/reports/privacy-online-report-congress/priv-23a.pdf
3. H.J. Smith, T. Dinev, H. Xu, Information privacy research: an interdisciplinary review. *MIS Q.* **35**(4) (2011)
4. R.A. Posner. The economics of privacy. Am. Econ. Rev. **71**(2) (1981). http://www.jstor.org/sta ble/1815754. accessed Nov. 11, 2021
5. P.M. Regan, *Legislating Privacy: Technology, Social Values, and Public Policy* (University of North Carolina Press, Chapel Hill, NC, 1995)
6. N.B. Rakshit, Right to constitutional remedy: significance of article 32. Econ. Political Wkly. **34**(34/35) (1999)
7. J. Mai, Three models of privacy. New perspectives on informational privacy. Nordicom Rev. **37**(special issue), 171–175 (2016)

Chapter 3
Data Security

One single vulnerability is all an attacker needs.—Windows Snyder

The rise in the volume of electronic digital data has given birth to concerns related to its legitimate use, archival, and security concerns. In the past, there were several attacks to encrypt, corrupt, or destroy data, including data breaches, data spills, and ransomware. Hence there is a need for mechanisms to prevent data from such attacks while keeping it available for its designated usage. Data security consists of mechanisms, principles, and policies to protect data and ensure confidentiality, integrity, and availability. The data needs to be protected from various threats of unauthorized access, resulting in data loss, data corruption, or data hostage. Data security ensures confidentiality by protecting the data from unauthorized access. Data security also enables data integrity by protecting data from unwanted modification and data erasure. Availability of data is ensured when data remains accessible for legitimate users. The data security mechanisms consist of various security controls, computer systems, and software; together, they ensure that services and information systems are available when needed. For example, data security in financial systems is securing the financial database access to prevent unauthorized access, regular data archival to ensure data is not lost or corrupted and keeping the database available for legitimate users all the time [1].

Some industries with confidential user data must comply with data security regulations, such as medical and financial firms. For example, health care firms in the USA must adhere to HIPPAA standards, whereas payment gateways or banks need to ensure secure storage and processing of payment data. Today, various public and private sector organizations apply data security practices to adhere to their social responsibilities of securing confidentiality, integrity, and availability of customers' data. Data security is vital for an organization's survival and can directly impact its critical assets and customers' private data.

The structure of this chapter is as follows. In Sect. 3.1 presents background of data security. In Sect. 3.2, we discuss the importance of data security, including data breach and data spill, cost associated with data breach, and various data security challenges.

© The Author(s), under exclusive license to Springer Nature Singapore Pte Ltd. 2022   41
S. Shukla et al., *Data Ethics and Challenges*, SpringerBriefs
in Computational Intelligence, https://doi.org/10.1007/978-981-19-0752-4_3

We discuss data security solutions and mechanisms in Sect. 3.3, including data-centric security controls, governance and compliance, and regulations. In Sect. 3.5, we discussed a famous case study on widespread ransomware and how it affected data security.

3.1   Understanding Data Security

There are various nuisances we need to understand first to appreciate data security. The background will help us understand the terminologies used in the security domain to better understand data security, threats, and mechanisms.

Unauthorized Access

Unauthorized access occurs when a person gets entry to data, software, system, or computer network without permission. When there is an access violation according to the security policy formed by the owner or operator of the system, we term it unauthorized access. Unauthorized access may still occur when a legitimate user accesses the information they are not entitled to. There are various ways to obtain authorized access, including password guessing, exploiting software vulnerabilities, or social engineering. Here, we are focusing on digital unauthorized access tactics [2].

Password guessing is the most common form of unauthorized access. The password guessing can be done manually by using social engineering, phishing, or carefully researching the person to derive a password based on the personal information. Hackers also use sophisticated softwares to guess passwords based on personal information such as a person's name, date of birth, or previous passwords.

A software error or bug is a mistake that causes the software to not work as per requirement. Sometimes the bugs are notable vulnerabilities, which attackers can exploit to gain unauthorized access to underlying data, networks, or operating systems. Attackers inject custom scripts or code into the software, and the software bug allows them control of the application. Once an attacker gets the control, the underlying data can be stolen, encrypted, or erased altogether.

Social engineering is about taking advantage of human vulnerabilities. It is widespread in the cyber world, for example, convincing a person to hand over their credentials, passwords, or any other sensitive data. These attacks usually include a form of psychological manipulation and employ malicious links in websites, pop-ups, emails, or text messages. Impersonation, ransomware, smishing, and phishing are typical social engineering techniques to gain unauthorized access.

Data Loss

Data loss happens when data is encrypted, corrupted, or deleted and becomes beyond recovery by humans or software. Sometimes, data loss occurs due to human error when information is accidentally deleted or corrupted or when physical damage occurs to data storage. However, data loss can also occur due to a targeted attack where a virus steals, encrypt, or delete the data. Most of the time, data recovery is impossible or requires lots of time and advanced skills. Data loss impacts the business in many ways including, destroying functionalities, making business operations stall or crawl, or causing damage to the reputation. Regular backups and robust data security mechanisms are the default choice for data loss prevention [3].

Computer Virus and Malware

Malware is a generic term to describe malicious software programs created to wreck or misuse any software program or network. A virus is a typical type of malware that automatically replicates itself by injecting its code into other programs on a system or a network. A system can receive a virus through a malicious link present in an email and invoked when the receiver clicks the link. The link injects the virus or malware into the system, replicating itself to damage or steal data. Anti-virus softwares are usually deployed to guard against malware and viruses.

Tokenization

Many data privacy and protection regulations converge on protecting user identities present within personally identifiable information (PII). Tokenization is one mechanism where PII information is replaced with a unique data token. The token is alphanumeric and used to separate identities (data subject) from the data. Even when data is leaked, the identities are not compromised because the PII information is already tokenized [4].

Data Security Versus Information Security

Information security is a broad term, and data security is a subset of it. Data security and information security are often interchangeably used, but they are not identical. Information security deals with safeguarding information from unauthorized access, usage, alteration, or erasure. Information security has a broader scope that comprises the end-to-end flow of information, including knowledge, processes, communication, user interface, and data storage. Data security is a subset of information security and deals with data in storage. Data security focuses on the physical safekeeping of data, encryption, and erasure, preventing unauthorized access, usage, and altercation of the stored data.

Historically, information security started when humans felt a need to keep things secret. They devised various safekeeping measures of physical paper-based files. When computers replaced physical files with digital ones, network security was devised to transport the information securely to keep the information secrecy. The rise of the Internet changed the quest to the next level when information was kept on the Internet and held by several vendors. Cybersecurity was devised to safeguard against various online threats on data present on the Internet. Thus, Information security is an umbrella term for security measures applied at different layers; each layer corresponds to a security type, starting from data security, network security, and cybersecurity [5].

Data Protection Versus Data Security

Data security is distinct from data protection. Data security is designed to prevent a malicious attack upon personal or enterprise data, whereas data protection is designed to restore the data if required. Typical data protection methods are regular backups and snapshots of the data taken and stored at different geo places. Data protection relates to the mechanism of securing copies of data to restore in case of data loss or data corruption. In contrast, data security concerns prevent unauthorized access and distribution of the data.

One of the critical elements of security design is its layered approach. Usually, security measured are applied at different layers; if a breach occurs at one layer, the other layer defense should safeguard the systems. If data security mechanisms fail, data protection mechanisms come as the last line of defense in the security strategy. This means that if the data security strategy fails, the data protection strategy will facilitate system recovery by restoring clean data copy.

For example, a successful ransomware attack would encrypt the organization's data and create a data hostage situation. In this case, IT admins can restore the data from backups as a data protection strategy. WannaCry, was one such ransomware that encrypts files and appends filenames with ".WCRY" file. The data on the user system remains unreadable until the files are decrypted. The attacker ransoms the user to shell out money to decrypt data or threaten to delete them forever. This attack is possible due to a lack of adequate data security measures to prevent unauthorized data access. If sufficient data protection measures are applied, the system recovery can happen using backups and remotely stored data snapshots. Hence, data protection mechanisms give an additional layer of security when the data backups and snapshots are stored remotely.

Data Security Versus Data Privacy Versus Data Protection

Data privacy deals with data collection, usage, and sharing in the light of keeping the data confidential. There is a substantial degree of commonality between data security and data privacy. For example, file encryption is a means to protect data privacy but is also used for data security. The principal distinction between data privacy and data security is that data security is about defending against malicious threats to data. In contrast, data privacy is about safeguarding the user identities present in the data. Data privacy ensures that information is available to the individuals or systems for whom access is granted. For example, data encryption makes the data private but not necessarily secure. The encrypted data can still be deleted or corrupted by an attacker violating data security principles.

Data privacy is a lot different from data protection. Data privacy is around safeguarding the data against unauthorized access, while data protection is about ensuring a system can restore its data in the event of data loss. Notwithstanding these distinctions, data privacy and data protection are often used together. A prime example is Backup tapes, which are generally encrypted to block unauthorized access to the data. There are many similarities between data protection, data security, and data privacy regarding regulatory compliance.

Data Security Versus Cybersecurity

Computer security, cybersecurity, or information technology security (IT security) is the protection of computer systems and networks from information disclosure, theft of, or damage to their hardware, software, or electronic data, as well as from the disruption or misdirection of the services they provide.

The field is becoming increasingly significant due to the continuously expanding reliance on computer systems, the Internet and wireless network standards such as Bluetooth and Wi-Fi, and due to the growth of "smart" devices, including smart-phones, televisions, and the various devices that constitute the "Internet of things." Cybersecurity is a significant challenge in the modern world due to its complexity.

There are three categories in which cybersecurity threats can be divided.

- Cybercrime: A single person or group mounting an attack to gain financially or cause disruption is called as Cybercrime.
- Cyber-attack: It typically involves targeted information gathering to fulfil motives of an specific organization.
- Cyberterrorism: Usually used to cause panic and fear, a broader cyber-attack is intended to cripple electronic systems at scale.

There are various ways a malicious actor gain access to the computer system. Some of the typical techniques used to threaten cybersecurity are:

Malware

Malware is a malicious program created by hackers or attackers to disrupt or gain access to a legitimate user's computer. Malware is often transmitted through a voluntary email attachment or genuine-looking download [1]. There are several types of malware, including:

- Virus: Virus is a self-replicating program with malicious code that attaches itself with every new and clean file to spread.
- Trojans: A malicious program disguised as genuine software. Attacker misleads users into accepting Trojans on their system to collect data and cause damage.
- Spyware: A malicious program that records user activities secretly and sends them to attackers. For example, spyware can secretly capture and send credit card details.
- Ransomware: A malware that encrypts down a user's files and later threatens to erase them until the user does not pay a ransom for decrypt.
- Adware: Advertising programs used to distribute malware.

SQL (Structured Language Query) Injection

SQL injection technique is used to steal or disrupt data from a database. Attackers find vulnerabilities with data-driven applications and execute malicious scripts directly onto the database via SQL statements revealing sensitive information. SQL injection is the most common data-stealing technique used to steal or disrupt data.

Phishing
Phishing is social engineering attack when an attacker targets victims to divulge sensitive information over a reply to an email that appears to be sent from a legitimate source. Phishing misleads victims into believing the information seeker is genuine and often seeks sensitive information urgently.

Man-in-the-Middle Attack
Man-in-the-middle attacks happen when attackers intercept data flowing between two individuals to steal information. For example, the attack can steal data passing between a user and website when logged in to an unsecured Wi-Fi.

Denial-of-Service Attack
A denial-of-service (DoS) attack is the most severe form of attack when an attacker stops a system or service from serving a legitimate request, often by bombarding the service with a high number of manually created requests. The DoS attack makes the system overwhelmed and inoperable [1].

Compliance and Regulations
Several countries have adopted stiff regulations and compliances to make organizations accountable for data security, privacy, and protection. For example, US Healthcare companies are bound by Health Insurance Portability and Accountability Act (HIPPA) to protect patient information; European Union companies are bound by General Data Protection Regulation (GDPR) EU law. GDPR, HIPAA, and related regulations outline requirements for organizations to assure data protection, security, and privacy [6].

Data Security Vulnerabilities, Threats, and Risk
A **data** security vulnerability is a flaw or weakness in the system that increases the likelihood of exploitation. A data security threat is an event that occurs when an existing vulnerability is exploited. Data security risk is the potential for destruction, damage, or data loss. Data security risk is a function of a data security threat taking advantage of data security vulnerabilities to cause damage, erasure, or data loss. For example, an information system could be at risk of losing data when a threat of hackers gaining access to the computer network exploiting a vulnerability of poorly formed system access control mechanisms.

The data security threats could be a service accessing unauthorized data, an attacker gaining access to resources behind a firewall, a cyclone wiping out a data center, or an employee leaking confidential customer information. It is an essential step in data security to identify, classify, and evaluate data security threats based on the damage potential. Technical data security threats are hacking, cracking, malware, and data leakage. Whereas non-technical data security threats are physical, environmental, insider threat, and social engineering [1].

3.2 The Importance of Securing Data

Data security is crucial across domains, be it medicine, finance, or any public and private organization. There are several repercussions for not handling data security maturely. First, an organization's lawful and moral commitment is to protect its customers' data from unauthorized access. For example, Financial firms' processing payments are subject to standards such as Payment Card Industry Data Security Standard (PCI DSS), which obligate them to put adequate measures to protect customer's data and transactions. Second, reputational threat in case of a data breach. When publicized, any high-profile data breach or hack can permanently damage an organization's reputation and business, not to mention the financial or logistic consequences of data loss. The cost of restoring and repairing damage caused by a data breach or loss can spell a doom day for the business. We will see the definition of a data breach, data leak spill, and data loss and compare them.

Data Breach
A security violation, where confidential, sensitive, and protected data is viewed, transmitted, copied, stolen, or used by unauthorized users, is called a data breach. There are other names for data breaches, such as data or information leak, data spill, or unintentional information disclosure. There are different examples of data breaches, including hackers penetrating an organization's database or accidentally data spilling, causing unauthorized access. The information in question can be sensitive and confidential information such as customer name, credit card, trade secrets, or information related to national security. There are many reasons attackers launch a data breach attack, such as for personal use or enmity, to gain political mileage, or part of organized crime to make money by keeping the data hostage. A data breach can cause a damaged reputation or loss of business for the target organization. A data breach is often treated as a breach of users' trust from the organization or service provider. The victim and its customers whose data is compromised may often suffer privacy of financial losses because of the data breach [3].

Data breach can expose several types of information, including:

Financial data: example, bank details, credit or debit card numbers, financial transaction details, invoices, tax forms, or financial statements.

Personal Health Information (PHI): Medical information collected by the health care providers contains vital information regarding a patient's health history, treatment, and ongoing health condition. A popular health care standard around lawful handling of personal health information in the USA is The Health Insurance Portability and Accountability Act (1996), also known as HIPAA. HIPPA defines PHI as "information that is created by a health care provider [and] relates to the past, present, or future physical or mental health or condition of any individual".

Personally Identifiable Information (PII): Information that can alone identify a person falls into this category, such as name, phone number, social security numbers, or address or combination.

Intellectual property: information around trade secrets, patents, customers list, and contracts.

Vulnerable or sensitive information: Usually, information sensitive to national security, military, or political character falls into this category. Examples include classified documents, agreements, protocols, or meeting recordings.

As per a data breach study, personally identifiable information (PII) is the most stolen data, followed by financial data. In 2014, hackers injected malware into Marriott International's systems, a leading hotel chain in the USA, but it was undetected until 2018. The malware exposed millions of guest's private data, such as their names, phone numbers, passport details, and credit card numbers. The attackers implanted a Remote Access Trojan into Marriott International's system carrying MimiKatz, a utility used for sniffing user credentials. Marriot's inadequate data security mechanisms were no match for the state-of-art attack and resulted in the high-profile data breach incident. The breach had a devastating personal impact on the concerned individuals as Millions of passport and credit card numbers were compromised. They widely suspected that an adequate data security measure and regular audits could have uncovered the data breach way sooner.

Data Leak or Data Spill

A data leak occurs when an unauthorized transmission of information happens from within an organization to outside or external receivers. The transmission of the information can happen physically or digitally. A data spill is also a data leak when the data is accidentally or deliberately exposed to an unauthorized person or environment. Mostly the data leak incidences happen on the Internet via file transfer or emails. Typically, it occurs when the email is sent to an unauthorized person by mistake or when confidential information is disclosed to a wrong request. The majority of the data leak incidents are unintended and non-malice [3].

In February 2021, Dutch Municipal Health Services (GGD), coordinating Covid-19 testing and vaccination, confirmed a data leak incidence. The data included personal data of Dutch citizens meant for efficient and effective Covid-19 vaccination program. Per RTL Nieuws, the Dutch news service claimed that criminals bribed GGD employees to get hold of the data. Because of the obsolete digital systems and inadequate access control, Dutch citizens' sensitive information was accessible to all GGD employees, resulting in a high-profile data leak. Stringent data security mechanisms and frequent security audits could have saved the data leak.

Figure 3.1 shows the critical difference between a data breach and a data leak is the origination. A data breach happens from outside-in, whereas a data leak happens from inside-out. Hence, GCD incidence classifies as data leak, whereas the Marriot International case classifies for a data breach. Today, no perfect solution restores data entirely after the data loss caused by a data breach or leak. However, implementing adequate data security mechanisms and state-of-art cybersecurity practices can significantly reduce the threat of data breaches or leaks.

Data Breach Costs

The data breach cost can wreck the organization. These costs include the expenditures of identifying, acknowledging, and reacting to the breach, the downtime cost and revenue loss, and damage to the reputation and brand of the organization. In 2020,

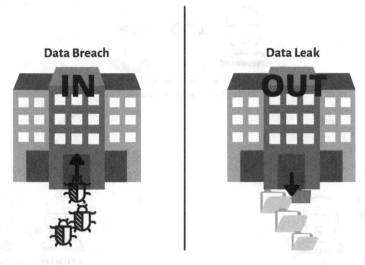

Fig. 3.1 Data breach versus data leak

the USA alone saw an average data breach cost of 8.64 million. The cost can have short-term and long-term impacts on the organization and its users.

The immediate impact of a data breach is disruptions to operation when data needs to be restored and operations need to be stabilized. The organization must notify its customers and minimize the impact of a data breach. Public relation cost goes up to give authentic and up-to-date information on the data breach impact and revival. The business needs to investigate the cybersecurity threat and vulnerabilities associated with the breach. A local government may fine the organization for not adhering to data security guidelines and regulations. Usually, data breach incidents result in declining stock prices if the organization is a publicly-traded company. Though, there is a far more lasting and devastating impact of a data breach in the long term for an organization. Customer's and partner's trust is reduced or completely lost post the data breach. The trust and sentiments issues with the organization resulted in a tarnished or damaged reputation and impacted the overall brand value. The organization may lose its existing customer base or find it difficult to onboard new customers. Another impact of a data breach is on the insurance premium paid by the organization since now insurers will find the organization riskier to insure [7].

Data Breach Cycle
Typically, an attacker follows the below steps to mount a data breach attack until they get access to the sensitive data they are looking for Fig. 3.2.

Reconnaissance—first, an attacker starts by selecting probable targets such as systems, protocol, or ports that are easy to access, penetrate, or compromise. Another technique the attacker can use is to mount a social engineering attack against a person having the required privileges to access the system.

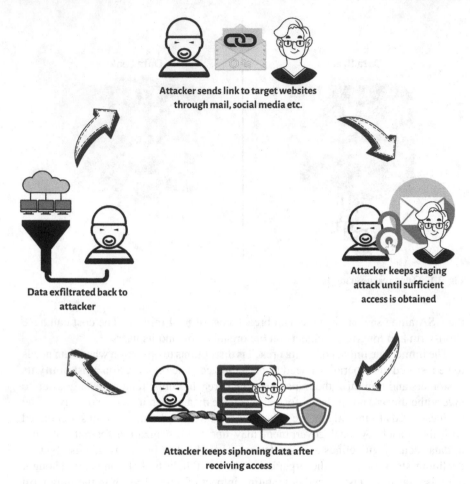

Attacker sends link to target websites
through mail, social media etc.

Attacker keeps staging
attack until sufficient
access is obtained

Data exfiltrated back to
attacker

Attacker keeps siphoning data after
receiving access

Fig. 3.2 Data breach cycle

Intrusion—the attacker attains the foothold in the system or network by breaching its security techniques.

Privilege escalation—the attacker might not access the sensitive data immediately after breaching the security. The attacker will escalate his permission either by impersonation or by running malware. The escalation will continue until the attacker can compromise the user's account and get hold of the desired data.

Exfiltration—Once the attacker gets the desired data, they will transfer it outside the organization's security perimeter and use it for personal gain, take data hostage, or resell in the black market [3].

3.3 Data Security Solutions

There are various practices and techniques to improve data security. Organizations adopt a mix of these techniques based on the data security risk, as one technique can not solve all the problems. By combining several techniques, organizations can considerably improve their data security posture. In this section, we will see various data security techniques.

Data-Centric Security

Data-centric security affirms securing data itself rather than securing port, network, system, or software. Data-centric security is rapidly emerging as the business has become more data-driven, and big data systems are now mainstream. More and more enterprises now depend upon digital information to run their business smoothly. The data-driven application's business strategy revolves around data, and data-centric security ensures that security is applied at the core of the business, that is, data [4]. The data-centric security model is inclusive of the following steps (Fig. 3.3).

- Discover: capability of knowing about location and nature of data, including sensitive information.
- Manage: the capability to specify policies around data access, modification, and erasure.
- Protect: the capability to safeguard against data loss, unauthorized access and use of data, or unauthorized data transfer.
- Monitor: consistent monitoring of data usage to determine legitimate use versus malicious use.

Technically, implementation of data-centric security relies on practices that make data self-describing and self-defending and implementing policies and mechanisms as per business context agnostic of data management technology. Data-centric security enables a smooth flow of protected data between application and storage.

Types of Data-Centric Security Controls

Data Discovery and Classification

The first step towards data-centric security is to discover and classify the data in question. Data discovery is often the first step in implementing data security models. It covers the organization's data kept across several platforms, such as the cloud and intranet. Data discovery and classification is an automated process that categorizes

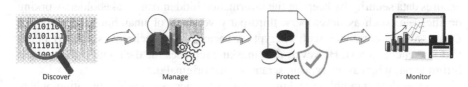

Fig. 3.3 Data-centric security model execution steps

the data based on the importance and suitable protection mechanisms to be applied later. Data security and protection mechanisms vary as per the nature and type of data; for example, data security and protection applied for credit card information of employees will differ significantly from the organization's patent and IP protection [4].

Identity and Access Management
One of the vital data security controls to ensure data-centric security is managing user identities and efficiently accessing permission. There are various techniques through which it can be implemented.

Identity and Access Management (IAM) is a method, business process, and framework that manages digital identities in an organization. IAM solutions allow IT administrators to create, modify, or delete user identities in the system and enable them to manage granular access permission for data within an organization. The authentication system is strengthened in various ways, such as single-sign-on, two-factor, or multi-factor authentication. These technical frameworks facilitate the organization to process digital identities securely. IAM implemented as policies support governance of user-profiles and management of access policies applied at a granular level in the IT infrastructure, including data.

Another control to implement data-centric security is using data access control to restrict access of data to selective users, groups, or roles. The access permission could be different for viewing, editing, or deleting data. The challenge with data access control lies with a thoughtful process to define the access controls according to data, importance, location, and who should have access to it and till when.

Data Encryption, Data Masking, and Tokenization
Data encryption is the most popular type of data-centric security mechanism. Data is converted from readable format to an unreadable or encrypted format using cryptographic techniques, for example, public-key cryptography. We convert data to encrypted format using a cryptographic algorithm by supplying an encryption key. We can only decrypt the encrypted data using a decryption key. Encryption makes the data secure from unauthorized access since the attacker needs a valid decryption key to decrypt the data and access sensitive information. Many organizations made data encryption a compliance standard to ensure data remains encrypted at rest (in storage) or in transit (send over the Internet).

Data Masking is hiding sensitive information by converting plaintext into non-recoverable encoded text. There are various techniques to achieve data masking, such as obscuring data dynamically, hashing, or duplicating the data. Data masking ensures data security by keeping the information hidden from stakeholders working on the data, such as developers, third-party vendors, or unauthorized personnel. Data masking correlates with encryption because both techniques convert plaintext into encoded text. However, data masking replaces the data with proxy or fake information, whereas encryption scrambles the original data.

Another data scrambling technique similar to encryption is Tokenization, where the sensitive data is replaced with a random string as a token generated dynamically from a token generation system. In order to de-tokenize the data, the same token generation systems is needed. The Tokenization also hides the data and protects it from unauthorized access. The token generation and de-token system should be logically isolated from the data processing systems and the applications to avoid attacks, such as token mapping table exposure, side-channel analysis, or direct attack [4].

Deletions and Erasure
It is a common myth that data, once deleted, cannot be recovered. In reality, sophisticated recovery software can recover data from media and storage, which a user has accidentally or intentionally deleted. In 2019, a lab investigation revealed that 25% of mobile devices were disposed of after formatting the media or deleting the files. They were unaware of safe data erasure techniques and therefore put their data at risk of a data breach. Data erasure offers the most productive and safest ways to wipe out data by overwriting the memory location with a gibberish binary pattern and making it unrecoverable by any data recovery software.

Backups and Recovery
The risk of a data breach always exists regardless of how strongly the organizations have applied the other data security mechanisms. Organizations need to stitch together data protection and loss prevention strategy to prevent data from falling into the wrong hands. It covers several types of protections.

Backup and recovery is a data protection mechanism where multiple duplicate copies of the data are created at regular intervals and saved at different geographic locations as backups. In data loss or data hostage, the data can be recovered using the latest backup available. Backup and recovery is a widely used data-centric technique used as the last data security measure when all other mechanisms fail. There are various out-of-the-box data backup and recovery software available which various options such as frequency, location, and type of backup and snapshots and their fast and efficient recovery.

3.4 Governance and Compliance

Regular data governance and compliance management form the essential part of data security strategy as they can help in improving data security. To stay compliant, organizations must abide by national, international, and industry-unique data security regulations. Non-compliance attracts heavy penalties and fees. Data security compliance aims to enable organizations to achieve integrity, availability, and data security. They serve a set of rules and guidelines to protect data from potential security risks. It comprises the following key elements:

- **Governance**: create strategies, controls, and policies imposed across an organization to safeguard data protection.

- **Risk**: identify and access potential data security threats and safeguard against it.
- **Compliance**: Ensure businesses practice data security policies according to the rules and regulations defined by industry standards when accessing or processing data.

General Data Protection Regulation

The European Union brought the General Data Protection Regulation (**GDPR**) to strengthen and unify personal data protection regulation. May 25, 2018, it came into effect with an exhaustive and precise set of rules and guidelines that recognize the idea that different levels of protection are needed for different "flavors" of personal data. It rated the highest level of protection for personal information such as genetic, biometric, or criminal history as the highest. Likewise, if the organization process a very high volume of data, they need to register with Data Protection Authorities appointed by the government. GDPR applies to all companies in Europe processing the personal data of European citizens. GDPR [8] imposes stiff fines, penalties, or sanctions on non-compliant organizations, making it one of the strictest regulations on data security in Europe. Some of its salient features are:

- **Consent**: Organizations need consent from individuals before collecting their data. The type of personal data collected defines consent decree.
- **Data minimization**: To minimize data collection, GDPR clearly defines rules for collecting personal data linked to a well-defined business use case. Non-compliance occurs if an organization uses the data for any purpose other than the one defined in consent.
- **Individual rights**: Data subject (the person whose data personal data is being collected) has a right to know the purpose of their personal data collection. They also have a right to be forgotten or erased. Additionally, the data subject has the right to be notified in the event of a data breach when their data is at risk of exposure.

The California Consumer Privacy Act

The California Consumer Privacy Act (**CCPA**) concentrates on consumer privacy rights. Since January 1, 2020, and imposed by the Attorney General of California, CCPA regulates personal computer data, such as IP addresses, cookies, biometric data, and data generated by IoT devices. CCPA applies to all the organizing, processing, and collecting of California citizen data. CCPA gives the right to an individual to know and understand the personal data collected and its usage [8]. The data subject has the right to opt-out of the collection or request for their data's deletion with no degradation of the service offered by the service provider. Like GDPR, consumers too have the right to be notified about data breaches and privacy failures. Non-adherence attracts the penalty of up $7500 per violation; in addition, they consume too may sue the companies for the data breaches up to $750 per record.

Personal Information Protection and Electronic Documents

The Personal Information Protection and Electronic Documents Act (**PIPEDA**), which became effective on April 13, 2000, is the Canadian privacy law built to establish trust in electronic commerce for private sector companies in Canada that processes consumer data. Apart from Canadian private sector firms, it also brings international companies that focus on Canadian customers. The regulation applies to personal data such as name, address, age, marital status, medical history, political alignments, and personal opinions. It stresses companies' consent from Canadian customers before processing their personal information. In addition, they must endorse transparent personal data practices and minimize data collection for specific purposes. Consumers have the right to access the data and assert its truthfulness. PIPEDA also makes organizations liable for any data breach or leak. Like GDPR, PIPEDA also makes organizations obligated to notify the affected individuals in data breach or theft. Non-compliance to PIPEDA attracts fines up to CAD$100,000.

Health Insurance Portability and Accountability Act

To protect the security of patient medical records and information, the US came up with Health Insurance Portability and Accountability Act (**HIPAA**) in 1996 [6]. The act enabled enhanced and efficient patient health care by providing secure healthcare information flow. The HIPAA act brought all US healthcare systems under the compliance net and mandated data security and privacy standards. The compliance net covers healthcare plans, providers, and clearinghouses dealing with healthcare information in written, oral, or digital form [9]. Secure healthcare information comprises protected health information or PHI, such as a patient's name, address, physical or mental condition, treatment provided, or fee for the provision of healthcare. Since April 2003, HIPAA privacy rule has been mandatory for the healthcare system in the USA, whereas the HIPAA Security and Enforcement rule has been compulsory since April 2005.

Payment Card Industry Data Security Standards

Major credit card companies have come forward to secure their customers' sensitive data and built a Payment Card Industry Data Security Standards (**PCI-DSS**). The compliance requires contractual commitment to be followed by credit card companies. A non-government global body named Payment Card Industry Security Standards Council (PCI SSC) have standardized the technical and operations requirements to be followed by the credit card companies. The standard covers all the entities which manage, process, or consume payment information (Table 3.1).

3.5 Data Security Audits and Frameworks

In order to comply with data security and privacy regulation, the organization must periodically perform privacy and security audits. The audit can help them uncover risks, vulnerabilities, and gaps with their security posture. Ideally, the organization

Table 3.1 Data security regulation

Data security regulations	Affected parties	Protected data	Prime obligations	Penalties/fines
Payment card industry data security standard (PCI DSS)	Businesses that store, transmit, or process credit card data	Credit card and payment information are digital and paper during transmission and storage	Develop a secure network and periodically test and monitor the system against security best practices Adhere to solid policies on access management around credit card holder's information Develop a program to manage data security risk and vulnerability of the system	Non-compliance will attract a fine of up to $100,000 per month Suspension of card acceptance
Health Insurance Portability and Accountability Act (HIPAA)	All healthcare providers in the US who collect, maintain, access healthcare records in digital or electronic form	Personally identifiable electronic health information (ePHI)	Protect the integrity, confidentiality, and availability of ePHI Identify and secure against likely data security threats, data leaks, or spills	Non-compliance will attract a fine of up to $50,000 per violation; the annual max is $1.5 million Can also attract up to 10 years of imprisonment
General data protection regulation (GDPR)	All businesses that process EU resident's data	Personal data of EU residents	Develop and maintain data security best practices to ensure secure processing of EU residents' data and safeguard it against unauthorized access, loss, or damage	Non-compliance will attract a fine of 4% of the organization's world-wide turnover or €20 million, whichever is higher

should choose a third-party and independent security expert firm to certify the security audit. The audit involves preparing a threat model and launching an artificial attack on the system, such as penetration testing against public and internal APIs exposed by the system.

To enhance data security and safeguard regulatory compliance, organizations need to align their security-based initiatives with a well-tested and proven framework. The security frameworks are developed according to best practices, research, and proven experience.

These frameworks provide repeatable, well-tested, proven procedures across organizations to adhere to data security standards. Below are some popular frameworks, such as NIST Cybersecurity Framework, ISO 27000 series, and BS 10012. NIST Cybersecurity Framework is custom-made for guiding organizations on best practices and standards around cybersecurity and information security. HIPAA-covered organizations can draw parallels between data handling practices guided by the NIST framework to secure PHI by safeguarding their cybersecurity posture. ISO 27000 series enables organizations to secure their employee, financial, and intellectual property data. It is a proven international standard for improving information security by helping the organization in preparing clear, concise, and comprehensive data security policies. BS10012 framework helps organizations comply with GDPR requirements, especially around data privacy.

3.6 Case Study

WannaCry Ransomware
On May 2017, a world-wide ransomware attack impacted computers running windows operating system by encrypting files across machines and demanded ransom in Bitcoin cryptocurrency to decrypt the files. It was a warm classified as ransomware cryptoworm. The ransomware was named after the wannacry worm, it is also known as WannaCrypt, WanaCrypt0r, WCrypt, or WCRY. It is estimated that the attack affected more than 200,000 computers across 150 countries and caused damage worth millions of billions of dollars. The worm is a network worm that spreads itself through other computers by replicating itself using transport mechanisms embedded in itself. The transport mechanisms first scan the systems for vulnerability using EternalBlue to access the computer. The worm then installed and executed itself using a backdoor tool named DoublePular.

It all started when US National Security Agency (NSA) identified an exploit with Microsoft's server message block (SMB) implementation and decided to use it for its offensive programs; they named it EternalBlue. However, the vulnerability got leaked or stolen by a hacker group named Shadow Brokers and used by wannacry developers to exploit and propagate a worm they created using Microsoft Visual C++ 6.0 framework. However, Microsoft detected the vulnerability on their own and released a security patch on March 14, 2017, for various flavors of Windows OS such as Windows Vista, Windows 7, Windows 8.1, Windows 10. The worm came with a kill switch, where first it tries connecting a few chosen DNS, and it replicates and executes itself only when the connection attempt to those DNS fails. Once in execution, it was encrypting all the files in the computer and running another executable to change desktop background showing the worm warning and ransom message and a timer. It will start replicating itself to the local network and do massive IP scanning to find vulnerable computes over the Internet, causing heavy internet traffic load from the affected system.

Though Microsoft released the patch in March 2017, many computers did not apply the security fix and remained vulnerable. It was only after two months when the worm attack was launched, and those unpatched computers across the globe were impacted.

The worm denotes the importance of data security practices at various stages. First, EternalBlue could have been kept under tighter access control where the data could not have leaked beyond the organization. Though Microsoft security patches are auto-updated by default, some organizations keep the security updates and are pushed only after their validations. Not keeping the operating system up to date with security patches makes the computers vulnerable to data security threats. Another and the most important learning from WannaCry ransomware was effective Backup and Recovery options. The organization must spend effectively to back up their most essential data periodically and restore it quickly. In the event of data hostage or ransomware, organizations can minimize the impact by quickly restoring the most recent data from earlier backups. The organization should also periodically audit its systems to find vulnerabilities and proactively fix them.

New data security attack variants pop up periodically. Like wannacry, other run-of-the-mill ransomware attacks (Petya/Nopetaya) were created to exploit similar vulnerabilities. The security expert has repeatedly suggested that safeguarding against a particular attack or vulnerability is not going to scale. There is a need for strategic investment into solid and proven data security and cybersecurity practices.

Reflective Questions

- Define Data Security and why it is essential for organizations?
- What are the similarities and differences between data security and cyber security?
- Elaborate on the difference between data security and data privacy?
- Give detailed reasoning on data security is the first step towards achieving data privacy.
- Write down differences among risk, vulnerability, and threat in data security context? Explain with a real life example of any recent data security risk.
- What is data-centric security? Elaborate on data centric security model steps? Why is it required to monitor data-centric security models post implementation?
- Define and elaborate typical data-centric security controls?
- Explain how does better identity and access management strengthen data security posture?
- Explain the importance of backup and recovery from data security perspective.
- What are the popular compliance and regulations for data security across the world?

- Explain General Data Protection Regulation from data security perspective, where is it applicable, what are the primary obligation it specifies, and what the fines and penalties it imposes for non-compliance?
- Write an essay on a recent data security exploit you are known of.

References

1. C.P. Pfleeger, S.L. Pfleeger, *Security in Computing*, 4th edn (2002)
2. C.P. Pfleeger, S.L. Pfleeger, M. John, Security in computing, fifth edition. *Computers & Security* (2015)
3. K. Fowler, Data breach preparation and response: breaches are certain, impact is not (2016), p. 240
4. W. Aaron, Enterprise Security: A Data-Centric Approach to Securing the Enterprise (2013), p. 324. Accessed: Dec. 27, 2021. [Online]. Available: https://www.oreilly.com/library/view/enterprise-security-a/9781849685962/
5. Cybersecurity vs. Information Security vs. Network Security. https://onlinedegrees.sandiego.edu/cyber-security-information-security-network-security/. accessed Dec. 27, 2021
6. Health Insurance Portability and Accountability Act of 1996 (HIPAA)|CDC. https://www.cdc.gov/phlp/publications/topic/hipaa.html. Accessed Dec. 27, 2021
7. Data of thousands of Dutch citizens leaked from government Covid-19 systems. https://www.computerweekly.com/news/252495983/Data-of-thousands-of-Dutch-citizens-leaked-from-government-Covid-19-systems. Accessed Dec. 27, 2021
8. R. Layton, S. Elaluf-Calderwood, A social economic analysis of the impact of GDPR on security and privacy practices, in *2019 12th CMI Conference on Cybersecurity and Privacy, CMI 2019*, Nov. 2019. https://doi.org/10.1109/CMI48017.2019.8962288
9. K. Kalkan, F. Kwansa, C. Cobanoglu, Journal of Hospitality Financial Management Journal of Hospitality Financial Management The Professional Refereed Journal of the International Association of Hospitality Financial The Professional Refereed Journal of the International Association of Hospitality Financial Management Educators Management Educators. J. Hosp. Financ. Manag. **16**(2)

Chapter 4
Data Ethics

In truth, every man, as soon as he emerges from the animal existence of infancy and childhood—during which he lives by the pressure of those claims which are presented to him by his animal nature—every man who is awake to reasonable consciousness cannot fail to remark how the life about him renews itself, undestroyed, and steadfastly subordinate to one definite eternal law; and that he alone, self-recognised as a creature separate from the entire universe, is condemned to death, to disappearance in unbounded space and limitless time, and to the painful consciousness of responsibility for his actions—a consciousness, so to say, that, having acted not well, he might have acted better. And, with this understanding, every reasoning man must stop, think, and ask himself—wherefore this momentary, indefinite, unstable existence within a universe uncompassed, eternal, firmly defined?"—Leo Tolstoy

Ethics is a branch of philosophy that deals with the moral behavior of individuals. It is a collection of principles that helps human beings to distinguish right from wrong. Most of the principles presented are also known as moral codes. Following the moral codes would make a person an ethical person. Moral codes are desirable only for the persons who agree with them. Being a social animal, a human being is part of some society. As it is impossible to exist away from a society of human beings, every human being is expected to adhere to the moral principles that are embraced by society.

The word "Ethics" is derived from the Greek word *Ethos*, meaning habit or custom. Ethics guides human beings to decide on right and wrong. Philosophers and thinkers have long been puzzled over this vital subject and have many insightful things to say. There are numerous philosophical thoughts on this topic. This chapter is about these deep philosophical questions and their practice in the realm of Data Science.

For many people, ethics is associated with moral principles. Most of the moral principles are emerging from the inborn wisdom of what is "good." As human beings grow up, inborn wisdom is perfected by various kinds of training. Protecting oneself from danger is one of the moral principles of every human being with rationality. From self-protection, one person moves to protect one's closest relatives, viz., children,

© The Author(s), under exclusive license to Springer Nature Singapore Pte Ltd. 2022
S. Shukla et al., *Data Ethics and Challenges*, SpringerBriefs
in Computational Intelligence, https://doi.org/10.1007/978-981-19-0752-4_4

Fig. 4.1 Human dilemma
over good and evil

parents, and siblings. From the closest relatives, one moves to protect other relatives, friends, etc. This is a process of prioritized protection. Protection one seeks or
provides is from all possible dangers or attacks. The greatest value of human beings
is life itself. Hence, protecting life is the primary duty and responsibility of individuals. The second most important value of human beings is reputation. In some cases,
reputation is weighed above life itself. If reputation is hampered, some people end
their lives themselves (Fig. 4.1).

For many, their moral principles are founded on religious teachings. Religions
play a crucial role in the formation of ethical human beings. Even though people
may follow different beliefs, most beliefs agree on several ethical principles. Some
of them are: do good, avoid evil, do not steal, respect elders, do not cheat, care for the
parents, provide for children, etc. When people come together and form organizations
and communities, they carry forward their collective and leading value systems. Most
often, the institutions created by human beings follow ethical principles based on their
leadership and sometimes based on the majority of the members of the institutions.
Business institutions or companies follow ethical principles mostly based on their
foundational values of them. Companies are not human beings, and therefore they
can not act independently. The morality of a company or an institution is that of the

members of it. Hence, it is not straightforward to attribute morality to companies. Yet, it is expected for companies to act ethically.

Evaluating a decision or a situation to be right or wrong is not very easy. Although some decisions and situations are straight and explicit, in many cases, right or wrong is determined by the expert analysis of the decision or situation by people of authority. This is the reason why most institutions and societies have established justice maintenance systems. They make sure that ethical principles are not violated.

Ethics has cultural implications as well. Customs and traditions change from society to society and from culture to culture. Marriage, relationship, ownership, etc. have deeper cultural significance. There are societies where meeting people of the opposite gender before marriage is not very encouraged. In some countries and societies, sharing the same apartment or accommodation with young students of the opposite gender is forbidden. If it happens, it is treated as immoral behavior. In some societies, women are supposed to wear certain dresses. Exposing parts of the body, body contacts, etc. are treated as immoral and people are classified based on that.

It is alright in some societies when people in authority break the law. For the sake of maintaining the safe passage of the so-called very important persons (VIPs), priorities and traffic rules could be changed. People give concessions to some of the ethical violations when influential people are involved in them. Although many cultures and societies do not encourage 'late entry' for any function, it is normative for people of higher authority to be late. When the poor need to be in queues, the rich are given easy access. In the vaccination scheme adopted with the COVID-19 pandemic, many countries prioritized it not only for the healthcare workers and people with vulnerabilities but also for economically and politically influential people.

According to Epley and Kumar, ethics has to be designed in the context of any system rather than taking it for granted or considering it as assumed. Organizations and individuals have to consider ethics as part of the fundamental design of their activity. This is on the ground that neither an organization nor an individual is totally good or totally bad. If founded and nurtured based on the publicly accepted pillars of ethical principles, developing and maintaining ethical culture is not a difficult task.

In this chapter, we will discuss various types of ethics. However, the primary focus of the chapter is ethics in the realm of Data Science. As a fledgling field of academic discipline and business endeavor, ethical foundations would make it integrated with the ethical order in the world. Data, in particular Big Data, has an unimaginable effect and impact on society. In every walk of human life, data is discussed and used or misused. Unless we insist on high standards of ethical principles in the collection, storage, handling, and transfer of data, Data Science processes will have unprecedented repercussions in the world. Companies and highly individuals have already been using data heavily.

Business decisions are taken based on having an enormous effect on society, and we are still trying to understand this impression. As in any human endeavor, we need to insist that Data Science processes also follow high standards of ethical principles. It must be addressed consciously beyond the broad areas where it is easy for data scientists and the immediate beneficiaries to reach a consensus about right and wrong. This chapter will also discuss the basic norms of "socially acceptable behavior" for

Data Science Practitioners. There will be issues where it is not so easy to agree about what is right. The main goal is not to focus on these challenges but on the numerous issues to reach a standard agreement.

The structure of this chapter is as follows. In Sect. 4.1, there is a detailed discussion on the three types of ethics. In Sect. 4.2, individual views on ethics are reflected upon. Ethics in the domain of Data Science is presented in Sect. 4.3. A brief presentation of data ownership is given in the next section. Sections 4.5 and 4.6 are about Informed Consent and Institutional Review Board. In the next section Data Repurposing and Permissible Purpose are explained briefly. In Sect. 4.8, some popular cases of ethical violations with the help of data manipulation or repurposing are given.

4.1 Types of Ethics

Ethics is a body of written and unwritten principles and guidelines of moral behavior. However, ethics cannot be restricted to written laws or customs and practices alone. A person or an institution with ethical principles respects the written codes and then transforms them into the spirit of the codes. Ethics that are in the form of precepts and rescripts elevate an ethician and an ethical person to transform human existence. Ethics could be addressed and explored from different perspectives. In the current context of ethics associated with Data Science, we look at applied ethics with the underlying stress on meta-ethics and normative ethics. In the realm of applied ethics, we can look at ethics from three different angles. They are personal ethics, common ethics, and professional ethics. They are not exhaustive and exclusive classifications of ethics. Nevertheless, the approach of personal and common ethics clubbed with professional ethics can answer many questions related to ethics in the fledgling science of data. In data science, persons and society are involved. At the same time, as an emerging field of super-utility, it is part of the professional life of all forms of human existence.

Exis wider than what is stated by law, traditions, and public opinion. In professional life, human beings deal with one's personal needs, goodwill of the society, and the responsible and accountable profession. Hence, ethics can be understood from these three perspectives, viz., personal ethics, common/societal ethics, and professional ethics (Fig. 4.2).

Personal Ethics
Philosophically, the word person is understood as an embodiment of individual existence with a rational nature. This Boethian construct of the concept of person gives stress to the rational nature of a human being. Many philosophers consider human beings as rational beings. The rationality or the rational nature of a person is expressed through the effective use of the choices one makes. Many philosophical schools name this unique characteristic of human beings as 'free will.' A being endowed with free will, or a person who has choices at her/his disposal is the one who is connected to personal ethics.

Fig. 4.2 Types of ethics

For every action, human beings have a personal choice to make before it is taken. Human beings are made of these choices. An ethical person is a person who takes ethical decision-making. It is she/he who is made up of a continuous chain of ethical choices. The goal of personal ethics is not to create a bunch of conformists. History has numerous examples of individuals rising above the customs and traditions of societies to lead the societies to become ethically sound. There are more than a thousand species that are cannibals. Human cannibalism was universal. Globally, human cannibalism is highly condemned. Societies find it as one of the severest of ethical violations. It is abolished and condemned by societies because of the individual leaderships across the cultures.

Several crimes of the society such as mob lynching, war, and war-related atrocities were objected to by individuals. Individuals had raised their voices against these organized crimes even at the cost of their lives. Once an individual uses her/his moral voice to rise against social evils for all sorts, society gives in. It is the role of ethical persons to lead crusades against social and individual evils. Even in modern societies, evils such as social discrimination, racial abuse, pogrom, monopoly, cultural exclusion, various forms of slavery, intolerance, denial of human rights, etc. have no dearth.

In the world of data, all the above evils have their manifestations. Data fragmentation, data manipulation, selective misrepresentation, data-filtering, online news force-feeds, cyber-bullying, trolling, character assassination, cyber-spying, e-pronography, spamming, etc. are not just individual evils. These and many other data-based evils of the cyber-world put the modern world into shame even in the middle of advancements in the rhetoric of human rights. Cyberworld gives false identities to desired individuals. Many use this anonymity in the information superhighway or in the deep-data-sea to channelize their perversions, insanities, and parochialisms.

In spite of personal interests and in the comfort of cyber-anonymity, an individual needs to make personal choices. Conflicts are going to exist forever. When it comes to conflicts of interests, what an individual decides to choose makes the cyber-world just or unjust. In the physical world anonymity is no more a possibility. Fake identities and multi-locationalies are almost impossible or extremely hard. In the virtual world, there are parallel worlds with fake identities and multi-locationalities with IP spoofing.

Hacking and manipulating data have newer and newer dimensions. Nevertheless, the ethical life of a person built on universal principles of the good life would be the most desirable thing even in the cyber world. Although it is built on personal choices, the choices of individuals in cyber-engagement are to be matched with the value systems of the society one lives in. A disagreement with the existing systems will create fissures in a person's relationship with oneself first and then with the fellow human beings and with the society at large. Personal ethics is all about following good societal norms and on top of it practicing altruism, virtues, and the sacrifices one makes for the betterment of oneself, ones' family, and the society at large while handling all forms of data.

Common Ethics

Common ethics is also understood as societal ethics. Common ethics has its source in personal ethics. In the physical world, the evolution of common ethics was mostly due to the existence of borders. The borders of countries, religions, races, organizations, etc. helped and forced ethical principles to come to effect when the majority of people agree on them. Common ethics is the ethics of everyone with their consent or dissent. As part of a society or organization of all forms, its members have to endorse them. Being part of a society or organization is the fact that its value systems are approved and endorsed by its members. Hence, we could reiterate that common ethics is nothing but personal ethics. On the contrary, personal ethics could be stronger than common ethics. Before even the societies had enacted norms and ethical guidelines, individuals practiced them. Hence, many philosophers argue there is no such ethics as common ethics.

However, it is evident that every society/organization has its written and/or unwritten rules. The Constitution is the best example of such a code. One has to necessarily respect and adhere to the constitution if one is part of a society/organization. In most cases, one has no choice in the case of common/societal ethics. Arriving at common/societal ethics is mostly by consultation, debate, and discussion. They are also practiced due to their existence of them as traditions (Fig. 4.3).

Professional Ethics

When it comes to professional ethics, it is very much context-oriented. Professional ethics is common ethics for the individuals involved in particular professions. It is applicable to all persons involved in that particular profession based on their roles. It is literally a role-specific set of ethical guidelines. Originated in the context of

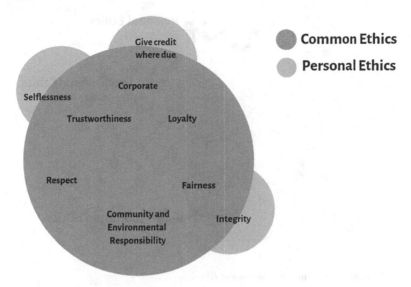

Fig. 4.3 Common points in personal ethics and common ethics

business, it is spread across all types of professions including that of the non-profit sector. Like any other form of ethics, professional ethics also has some universal values such as honesty and accountability, trust and loyalty, etc.

Apart from universal values, professional ethics demands ethical guidelines associated with a particular company or a profession. Even when a person is not working under an organization, professional ethics associated with the profession is binding. Lawyers, consultants, designers, physicians, writers, etc. follow the ethical and moral principles of their profession. When they work in the organization related to their work, additional ethical guidelines are imposed.

Practitioners of professional ethics add value to the profession and to the organization in the public sphere. This eventually builds confidence in the stakeholders of the organization. Professional ethics associated with judiciary, executive, and legislature create ethical nations. The origin and the existence of professional ethics are around the legal existence of the organization itself. Most of the professional ethics are time-bound. One is bound by professional ethics as long as one is under the legal contract of the employment of the organization. Similarly, one is bound by professional ethics until one practices that profession. Hence, the legal switching of profession or organization changes the boundedness of the professional ethics of the person (Fig. 4.4).

Professional ethics can have a conflict with personal ethics. Personal ethics may not endorse professional ethics too. Respecting privacy is a very important value. In the virtual world, a vast amount of data of customers is stored by companies. Customers could be the ones who use the gadgets produced by the companies or the ones who use their services. In some cases, companies collect data beyond the contractual requirements. Should an employee know this breach into the privacy

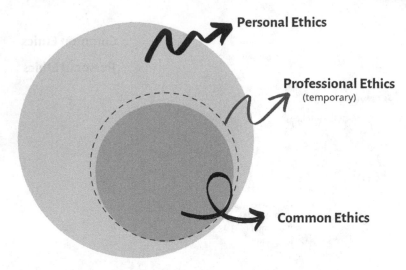

Fig. 4.4 The temporary nature of professional ethics

of individuals who continues to work in that organization? Many companies have established ethical principles. Unbeknown to the user, data is collected by many of them. Collected and stored data are in the hands of the employees. The sensitive data including passwords and pins, interests and disinterests, choices and decisions, etc. that are encrypted and unencrypted could be very vulnerable if they are stored beyond the need of the company. For example, should a hospital store the data beyond a certain time span in an encrypted manner? Should banks and financial institutions share the customers' information with marketing companies? In these situations, individuals should have the right to leave the organization honorably citing properly the conflicts of interests.

4.2 Individual View on Ethics

Ethics is founded on individual choices. Ethical principles are the vaccines for human existence for good. Hence, in most cases, individuals develop an affinity towards moral codes and principles. They are in agreement with the value systems. Although freedom is one of the most important values for any human being, in the context of a society or organization one has to sacrifice a huge amount of freedom. Small and big sacrifices help society become better. Hence, it is advisable for individuals to look for the common good before any individual decision. It has been said that individual ethics will be positively related to ethical intentions [1]. In the words of the famous English philosopher Thomas Hobbes, human beings agree to live together with a social contract for the expected return of cooperation, happiness, and peace. They become social beings because they use their rationality rather than living with individual selfish nature. Hence, ethics is one of the finest contributions of human beings individually and collectively.

4.3 Data Science Ethics

Some very important aspects of individual ethics are the protection of one's life and one's reputation. In both these cases, data plays a significant role. Data, as it is collected and stored, contains vital information of individuals, societies, and nations. Hence, if not carefully handled it can cause disasters. In the case of medical data and personal data, an individual's privacy is at stake if not handled carefully. From the points of collection, storing, and sharing, care must be taken.

In the sphere of data science, data is scientifically analyzed. Hence, vital information about a person or society is immediately at the fingertips of various people. If not protected, data could be used to malign the image of persons and societies. Analyzed data could be used for political and business gains. People with vested interest could use data scientists to achieve their personal goals. This might lead to widening the uneven structures prevailing in the society in terms of status, finance, and influence. The use and overuse of social media is a means of spreading manipulated data.

4.4 Data Ownership

The primary owner of data is the one who possesses it by virtue of being the source of the data. A collector of data is not the owner of data. Similarly, a keeper or distributor of data has only the delegated rights over data, if the delegation is done under a sound mind. A collector or keeper or distributor of data is only a secondary owner if permitted. The delegated power of the ownership of data has to be taken into consideration only in the particular context of transferring the rights of delegation. Data is to be handled with care while possessing and distributing. The ownership and control over data are not absolute in nature. Even the primary owner, the source of data, has only the right to use data ethically. Data ownership refers to both the possession of and responsibility for information. Ownership implies power as well as control. The control of information includes not just the ability to access, create, modify, package, derive benefit from, sell, or remove data, but also the right to assign these access privileges to others [2]. Various people involved in the responsible handling of data other than the creator data are the following.

- User or consumer of data. User or consumer possesses the right over data mostly on the exchange of an appropriate price.
- Enterprise—All data that enters the enterprise or is created within the enterprise is completely owned by the enterprise.
- A sponsor or Funder is the one who promotes the creation of data.

- Editor or cleaner or decoder or compiler has control over data as a technician. In many cases, data manipulation or fabrication can happen from these people.

4.5   Informed Consent

When one possesses data, one has every right to ethically use it or permit others to use it ethically. The relevance of informed consent is in permitting others to make use of the data mostly for a noble cause. A researcher or physician could gather the information from the owner of data so that the efficient use of data can be for the betterment of society. Informed consent is understood in multiple ways.

If a patient is informed of the impacts of a particular method of treatment, the physician or the responsible health care worker takes the consent from the patient. If the patient is not in a position to provide consent, an authorized legal person who is responsible for the patient gives the consent as delegated consent. Hence, minors, patients with unconscious state of mind or brain, etc. may not be in a position to give consent to the relevant persons.

In the case of research too, informed consent is required. Apart from informed consent, the health care worker or the researcher has to furnish the proper documents in justifying the need, relevance, and impact of the treatment or research.

The contract one signs, the license given, the copyright transferred, etc. are various other forms of informed consent. Although typically not treated as informed consent, contracts, licenses and rights are regular things in this era of mobile apps. While installing an app, the helplessness of the customer is made use of by the owners of the apps. Although they are supposed to collect on the minimum requirement for running the application, they collect an excessive amount of data. For example, if we install an application from the Add-ons of Google Docs or Google Sheets, the Add-on seeks in return complete access to the Gmail account. Further, in many cases, without even much relevance, Add-ons get unlimited access to camera, photos, and audio. Informed consent is not an absolute transfer of rights or consent. If it is expected so or if the owner of the data is blackmailed to do so, it is a serious ethical violation.

4.6   Institutional Review Board

When research is done on human subjects, data has to be collected, analyzed, and interpreted. In this process, the privacy of the human subjects should not be compromised. The rampant violation of human rights and inhuman human experiments conducted on the hostages, prisoners, enslaved people and the marginalized were the reasons behind the establishment of ethical guardians in research institutions. Most of these ethical guardians are named as Institutional Review Board (IRB). It is constituted based on international, national and institutional policies. Many funding agencies and publication houses seek the approval of the IRB for the funding and publication considerations.

Institutional Review Board is also known with different names with increased or decreased powers. The group that has been formally designated to review and monitor research involving human subjects. Some of those names are Institutional Research and Academic Integrity Panel, Ethics Committee, Research Ethics Boards, Independent Ethics Committee, and Ethics Review Board, etc.

The IRB reviews and analyses the proposed work taking into consideration the rights of the human beings who are supposed to be experimented, researched, or tested. If needed, IRB takes its own time to evaluate the proposal with the help of additional experts other than the members of the committee. In many cases, the IRB comes with guidelines based on the record of the reports of the IRB. It also takes into account the possible impact of the proposed research.

The genesis of the IRB is one of the most important developments in the field of research ethics. The dignity of human beings is upheld by the IRB if properly functioned. The IRB also provides rescripts to the researchers in the handling, analyzing, and presentation of the collected data from human beings. To make the work of the IRB serious, many institutions select independent persons with deep value systems and extensive research experience to be part of the IRB. In the modern research arena, the IRB functions as a conscience keeper and protector of human rights.

4.7 Data Repurposing

Data gathering is not accidental. Data is collected for the purposes that are agreed upon before the act of collection itself. To collect data from human beings, consent is required. The IRB is also involved in the approval of the collection of data from human beings. In many cases, after the collection of data, the researchers find new uses for it. This is in general forbidden. As we discussed earlier, various business houses and Apps also collect data in a general setting. Many cases of e-commerce companies massively selling data for business and political gains had surfaced in the last two decades.

Data repurposing [3] is the act of using the data beyond the original intention of collecting data. Mere reuse of data is not data repurposing. Salami slicing is an unethical practice in research publications that can come under data repurpose. To be ethically sound, when one wants to use the data for the purposes and the number of times other than the original intention of collecting data, consent, and approval from the appropriate persons has to be obtained.

Permissible Purpose
A service agency or a business has varieties of data of its customers. However, the data collected for the purpose of service or business does not make the service agency or business owner of it. The owner of the data is always the one whose data the agency of the business collected. Can a service agency or a business house

transfer the data it parks to a third party? Yes, if they have a 'permissible purpose.' Some of the permissible purposes are higher employment, background check, state demand, etc. It is highly desirable that in any such case of sharing the data with a third party, appropriate consent should be obtained by the service agency or the business institution.

4.8   Cases of Ethical Violations When Dealing with Data

Data-handling related unethical practices of companies that market apps and electronic gadgets.

- Transferring data to security and intelligence agencies.
- Supporting authoritarian governments.
- Using the data for the political and economic campaigns.
- Using data to suppress pro-democratic movements.
- Supporting and aiding anti-human rights programmes.
- Providing governments with Artificial Intelligence based infrastructure at cheaper rates for favors including tax reduction/exemption.
- International lobbying and opinion-generating activities.
- Research in E-spying and e-surveillance.
- Promotion of deepfakes, trolls, and memes.
- Encouraging cyber-stalking.
- Abetting monopoly.

Reflective Questions

- Have ethics anything to do with immediate gratification and delayed gratification?
- When personal ethics confronts common ethics, how does one proceed?
- Should professional ethics take precedence over personal ethics?
- Identify conglomerates and softwares/apps that are involved in each ethical violations mentioned in Sect. 4.8.
- Identify some cases where getting informed consent is nearly impossible.

References

1. H. Yu-Wei, The status of the individual in Chinese ethics, in *The Chinese Mind*. (University of Hawaii Press, 2021), pp. 307–322
2. D. Loshin, Knowledge Integrity: Data Ownership (Online) June 8, 2004 (2002). http://www.dat awarehouse.com/article/?articleid=3052
3. F.N. Doubal et al., Big data and data repurposing—using existing data to answer new questions in vascular dementia research. BMC Neurol. **17**, 1 72 (2017). https://doi.org/10.1186/s12883-017-0841-2

Chapter 5
Intellectual Property Right - Copyright

A wise man will always allow a fool to rob him of ideas without yelling "Thief." If he is wise he has not been impoverished. Nor has the fool been enriched. The thief flatters us by stealing. We flatter him by complaining.—Ben Hecht

Every human being is in possession of immense potential. One uses one's potential to optimize one's wealth, relations, and comfort. Although they are interrelated, one may or may not proceed from the other. The unboundedness of the potential can be seen in the productivity of human beings over the millennia in comparison with other living beings. All the other living beings maintain their status quo in their engagement with the world, with other living beings. Their past is the same and is without any record. Their future is predicted and is not planned.

When considering the case of human beings (*homo sapiens*), they have ever-evolving complex brain structures with unparalleled cognitive ability. They have memories that are episodic. They can create stories, evaluate the past, and plan for the future. They have self-awareness and formulate various theories of mind with a possibility of introspection and private thought. They have their own views of existence and have thoughts about higher beings. They are creative and ever-evolving in their brain process. They can possess things that are formed of matter. They also have matters for forms. In their capacity to possess matter and forms, humans also claim absolute control over certain things. One such thing is their cognitive capacity.

This poses a serious question. Do homo sapiens have absolute rights over their own intellect? They use their intellectual power to possess matter. There are rules and guidelines set by various societal structures on the limit of control and its usage. When it comes to the intellect itself, the debate over possession takes different turns. A material that is owned by a person is in the possession of that person according to the guidelines of society. This is the case with land, water, air, and other created materials such as money and everything that is purchasable by money. Human beings have an inherent feeling that whatever is produced by one person is one's own. However, as seen early, this ultra-selfish thought is not accepted by any modern society. Hence, what is human beings' right over one's own body and intellect, feelings, and dreams, creativity, and expression? Does one possess absolute rights over one's intellect? The answer to this question can not be understood with a direct answer. Hence, let us now consider the answer with some additional thoughts.

The ancient Greek thinker Parmenides is attributed to have stated that *ex nihilo nihil fit*. That is, 'nothing comes from nothing.' Although like any other organ of the human body, two brains may look alike, the output by each brain is different. Is it purely the contribution of the individual who possesses the brain? The training and nurturing of the brain and the information stored by the brain in the course of the growth of a human being different from that of other human beings. Taking into account all types of arguments one can easily arrive at the conclusion that the environment or the grooming of a person has a lot to do with intellectual performance too. Hence, 'my intellect is my property' may not be the right position that is to be taken by human beings. Nevertheless, human beings should be sufficiently acknowledged for their intellectual capacity as it is a natural property of them. Any infringement of intellectual property without attributing due credit in a timely manner is not acceptable in an ethical society. In the political economy, the property is considered private, public, or collective. Intellectual property broadly could be classified under private property. However, there are some schools of thought that put intellectual property under collective property also. In this chapter, we discuss various types of infringements of intellectual property in the context of copyright.

A property can be tangible or intangible. Intellectual property comes under tangible properties. The right over a particular property is the legal authority of a person to possess, modify and utilize the property. In 1995, the world trade organization (WTO) was successful in convincing its member nations to come to an agreement on Trade-Related Intellectual Property Rights (TRIPS). According to the TRIPS agreement, Intellectual Property Rights could be classified into several categories. They are Copyright, Industrial Design, Trademark, Trade Secret, Patent, Geographical Indication, and Integrated Chip Design. See Fig. 5.1. Violations with respect to copyright laws are mostly the same as the case of the other six categories of intellectual properties. We would be discussing only copyright in this chapter.

This chapter is presented in nine sections. After a brief description of intellectual property rights, Copyright is explained in Sect. 5.1. In Sect. 5.2, one of the serious concerns associated with the academic community all over the world, plagiarism, is explained. The connection between copyright infringement and plagiarism is discussed in Sect. 5.3. Section 5.4 is a brief description of the permissible limits of copying the copyrighted materials. Originality and creativity are defined and briefed in the next section. Section 5.6 deals with research and publication ethics. Some cases associated with intellectual property debate are given in Sect. 5.7.

5.1 Copyright

Copyright is the exclusive right granted to the author or creator of an original work. The rights are to copy, to distribute, and to adapt the work. Although a person owns the copyright to a creative original work that is presented in a tangible medium, the

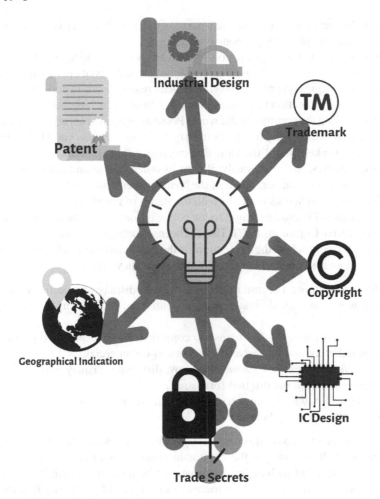

Fig. 5.1 TRIPS' classification of intellectual property

exclusiveness of the right is not permanently granted. Most of the countries in the world have enacted copyright laws. Copyright is vested with the creators until the last member of them and then a fixed number of years. The bonus time for the copyright in many countries ranges from 25 to 100 years.

Copyright is over material that is tangible. It is treated as data. They include audio-visual works, creative writings, visual arts, video games, softwares, plays, and musicals. All levels of co-creation are eligible for copyright. However, being a character or merely a performer will not give one right over creative work.

Creative work should not be an infringement of the privacy of others. If one needs to include a living person as part of creative work with all the explicit identifying information, if proper consent is not obtained, it will be treated as a violation of privacy. In such cases, the affected party could approach the law-enforcement agencies to take the proper legal course of action.

Copyright does not give protection to the common knowledge used in creative work. It also does not project facts, ideas, systems, or methods of operation. However, it protects the unique style and method in which creative work is presented.

Copyright is a protection to the idea conveyed through the creative work. It also is a way of rewarding the creators of the work. Possessing exclusive rights over creative work might lead to unhealthy trade practices or denial of access to knowledge. Hence, many modern thinkers are of the view that creative authors possess limited rights in the form of licenses. Simultaneously, it permits others to use and reuse the creative work in a responsible manner. Some of the creators even promote the idea of leaving every right over their works publicly with the stamp of copyleft.

Partial release of license and right to the public is done under the licensing scheme known as Creative Commons (CC). The CC licenses help the creators distribute their creative work freely while the protection of the work and the intellectual property rights are safeguarded. There are four CC licenses. They are:

- *CC BY (Attribution)*: License to copy, distribute, display and perform the copyrighted work and its derivative works with credit is given to the original creator.
- *CC NC (Non-Commercial)*: License to copy, distribute, display and perform the copyrighted work and its derivative works non-commercially.
- *CC ND (Non-Derivative)*: License to copy, distribute, display and perform the copyrighted work in the original form itself.
- *CC SA (Share Alike)*: License to distribute the derivative work under a license identical to the existing license.

The advantage of copyright is that it duly acknowledges the creator for a creative work produced. It encourages the creator to continue with creativity. The financial rewards associated with copyright protection help create a competitive spirit of promoting progress in society. It augments various types of prosperity in society.

Some databases also come under the privilege of protection like that of copyright. Databases are not creative works. They are creative compilations. Hence, a database cannot be compared with a creative work but for its 'originality.' The unique way in which a database is organized such that the storage, retrieval, modification, and deletion of data stored is faster and smarter demands a creative tag. Therefore, the owners of databases get their collection copyrighted even when the content is independently copyrighted or copylefted by the individual authors.

Copyright law is also there at the service of Technological Protection Measures (TPM). Technological Protection Measures are also known as Digital Rights Management (DRM). The purpose of TPM is to restrict and prohibit copyright infringement by way of controlling the way in which a work is used. TPM includes software, devices, and other technologies to control access to digital content or to prevent users from copying or sharing that content. With the ever-increasing traffic of digital content transactions over the internet, it becomes necessary for the individual users and creators of digital content to rely on TPM.

5.2 Plagiarism

Plagiarism originates from the Latin word *plagiarius* meaning kidnapper. Plagiarism is the act of scholarship-nabbing. It is the presentation of someone else's work or ideas as one's own. In any kind of creative work, the thumb rule is to give credit to the original creator for every part of one's work that is not one's own creation. If the timely acknowledgment is not given in a professional and proper manner, one can be accused of plagiarism. It will be treated as academic theft.

Most countries have come up with guidelines to curb plagiarism. The Tertiary Education Quality and Standards Agency (TEQSA) in Australia, The UK Research Integrity Office (UKRIO) in the UK, the University Grants Commission in India [1], the Office of Research Integrity in the US, etc. are some of the agencies that conduct regular campaigns against plagiarism.

Plagiarism is more than a similar to other works. Plagiarism occurs in all the fundamental aspects of creative work including the method, framework, language, and expression. Plagiarism can occur through four different ways, viz., intention, omission, commission, and ignorance. Accidental plagiarism is rarer than solar eclipse but plagiarism is an accident that cut-short one's professional life.

A severe case of plagiarism is verbatim lifting. It is the direct pasting of a work in another work for academic or industrial benefits. Even with attribution, verbatim copying is not acceptable. One needs to present the matter in one's own unique way and acknowledge the source in the style of referencing as desired by the work. With minor changes in the used words, some people present someone else's work with or without proper citation. This also comes under verbatim lifting. If there are no other ways to present something, i.e., poems, equations, formulas, etc., then verbatim copying is permissible. It is not difficult even for a basic plagiarism detecting software to identify verbatim lifted content.

One of the most disputed debated and misunderstood types of plagiarism is self-plagiarism. Does self-plagiarism exist? How is it condemned? How is it justified? Under copyright law one person has unlimited rights of copying, modify, and trading one's own work. If this is the case, then what is self-plagiarism. The problem of self-plagiarism is to be interpreted within the context of multiple benefits. Several copies of a book can be sold. It is justified. However, if one person recycles one's own work and presents it as a new work, then it is not ethically appropriate. In the cultural field, this may not be very significant. A refurbished dance or play or film might be better appealing to the public. However, in the academic world, if the work is not novel, original, and unique, it only burdens the students, teachers, researchers, and scientists. By repeating or using with minor modification introduction, literature review, conclusion, etc. in a new work of one's own, a researcher commits the ethical violation of self-plagiarism.

5.3   Copyright Infringement and Plagiarism

A creative work invests in an author's financial and non-financial rights. Among the non-financial rights, the most important one is a moral right. Moral rights bestow in the author the right to enjoy all different kinds of benefits of the creative work such as satisfaction, right to improvise the work, watching the progress of it, knowledge to interpret it, etc. Moral rights also come with responsibility. There are many instances where when one section of people can enjoy work, another section hates it and a third section is neutral about it. A picture drawn can bring revenue to the artist on one hand and on the other hand, it can begin communal riots. Hence, multiple aspects of moral rights have to be understood by the creator of the work and the owners of it.

The copyright holder has the right to use and distribute creative work. If someone uses and distributes a copyrighted work, it is known as copyright infringement. Copyright violations have no border limits now. Some violate the copyright given to the author of the work for the sake of money. Some violate the copyright rules because they do not endorse the concept of copyright. The reasons against copyright are digital freedom, freedom of information, the high price of the copyrighted material, monopoly, etc.

As seen above there are criticisms against the excessive control over creative works by the international and national copyright laws. Although it is not always justified, copyright infringement could be interpreted as a reaction against laws. Hence it is a statutory violation; a violation of the existing laws. Anyone who violates a law would invite penal sanction.

Although plagiarism is an academic theft, most countries in the world have not enacted it as an offense to criminal procedures. Plagiarism is considered as an ethical violation or precisely a moral corruption. There are procedures to deal with various types of plagiarism.

5.4   Permissible Limits of Copying

Copyright literally means right to copy. In the spirit of the law of copyright, it is established that copyrighted material is acknowledged as the work of its creator. Being a creative work involves so much effort from the side of the creator. It is natural that the worker demands his/her wage. This comes under natural justice. It is the natural justice of remuneration. As indicated earlier, there are financial and non-financial benefits to creative work. The primary beneficiary of both these benefits is the author of the work. If there are people to assist the author of the work, then they also need to be remunerated.

However, there are cases where every use of creative work needs to be compensated economically. When it comes to the context of personal study or research, most copyright laws permit the user to copy the matter. In all these cases, the condition of 'for private use only' is insisted upon. As far as the spirit of the copyright law is honored, a user may be permitted to use copyrighted material. Further, incidental copying of a work for another creative work is not a serious offense as long as the due credit is given to the original creator of the work. To be more specific, purposeful omission of the credit due to the creator of work is both a statutory violation and an ethical violation.

Is it fair to copy copyrighted material within the purview of the copyright laws? Fairness in the copying and using of copyrighted material depends on the nature, significance, effect, and purpose of copying. Further, if the original work is not affordable for a reasonable price for the party concerned, the seriousness of the offense would be diluted. Hence, the very act of copying a copyrighted work is not that unfair here.

Many countries permit library exceptions for copyright laws. Since the library is a public institution, limited copying can be permitted by the library for the purpose of private study and research. However, selling copied material for financial gains has to be fully avoided. Library exceptions change from country to country. Rare and unavailable books due to the non-availability of printed copies could be made available to the public by the library for the greater good of sharing knowledge. The same principle applies to books that are damaged, non-usable, or lost.

5.5 Originality and Creativity

Originality is the ability to generate a creative work that is not the reminder of another available work. The significance of originality is that human beings have the potency to think and create using a different permutation in the creative act. Music, play, dance, etc. are the best settings for original and creative work. If there is a measure for creativity, it is originality. Creativity is in its highest form when originality is given utmost importance.

As rational beings, human beings are the only living beings who can innovate. The degree of originality varies from person to person based on the nurturing and desire to innovate. For many, it is easy to imitate. Although creativity begins with imitation, imitation should be the nature of work throughout a human engagement. Research is one area where there are unlimited possibilities for innovation. Research without originality and innovation is called redundant research. Redundancy leads to duplication of research and resources. Copyright is a deterrent to being innovative and original.

5.6 Research and Publication Ethics

Research is one of the finest human endeavors. Human creativity, originality, and innovation have no better platform to perform than research. Human beings are fundamentally research-oriented. In the decisions taken every day, one takes sufficient care in protecting one's interest. This is possible only because of the research about the past and future that is done formally and informally by human beings. In everyday discussions, inquiries, observations, and experiences, humans record their learning. This learning could be used for one's own progress. If these learnings are presented on public platforms in the form of publications, it helps society at large. Research and publication ethics can be considered as part of professional ethics. Intellectual property is the most important tool that is used for research.

Once venturing into research, one has to look into certain ethical principles. Most important among them are respect, beneficence, and justice. Respect is to be given to the subjects of research and all people around the research ecosystem. If human beings are involved as the subjects of research their autonomy should be respected and the ones with diminished autonomy should be protected. When the research is about other living beings or inanimate things, laws of the land have to be adhered to. There should not be even a minor level of disrespect and misuse.

Even if the motivation of the research is extrinsic, beneficence should be the guiding principle behind every form of research. The world is better today with tech-nology and medical treatment only because of the research done by great researchers in the yesteryears. Thus, it is the responsibility of every human being to be involved in research with the faculty and facility one has. When involved in research whether theoretical or empirical, no harm should be done physically or emotionally to the subjects or to anyone. A researcher should be keeping in mind with utmost care and concert to maximize benefits for participants and minimize risks for participants.

From respect and beneficence emanate the principle of justice in research. Research is not for research sake. Thus discrimination, negligence, and prejudice should not find a place in research. Justice in research is also implemented by sufficiently acknowledging and rewarding the subjects of the research.

The publication is one of the most important follow-ups of research. After obtaining a research outcome, publish all truths that are charitable and relevant. The golden principle that is applicable to good publication is to cite everything that is not of the author and explain everything that is of the author. Although various countries and independent agencies have come with their guidelines for publications, every researcher and publisher should be committed to a higher level of publication ethics. The Committee on Publication Ethics (COPE) brings out ethical principles and ethical breaches from time to time.

5.7   Case Studies

- Copyright Protection for API
- Copyright Protection applicable to Database
- Technological Protection Measures.

Reflective Questions

- Should copyright be exclusive? Justify your answer.
- Identify various cases of permitted copying.
- How is originality determined?
- What are the ways to curb plagiarism?
- Should plagiarism be treated as a criminal offense?
- Is copyright infringement culture-dependent? Justify your answer.

Reference

1. V.K. Ahuja, *Intellectual Property Rights in India*, 2nd edn. (LexisNexis, 2015)

Printed in the United States
by Baker & Taylor Publisher Services